BREAKTHROUGH

BREAKTHROUGH

TUNNELLING THE CHANNEL

Derek Wilson

CENTURY

London Sydney Auckland Johannesburg

in association with

EUROTUNNEL

Text copyright © The Channel Tunnel Group Ltd 1991

For copyright details of all the illustrations see the picture credits

All rights reserved

The right of Derek Wilson to be identified as the author of this work has been asserted by him in accordance with the Copyright, Designs and Patents Act, 1988

First published, in association with Eurotunnel, in 1991
by Random Century Group Ltd, Random Century House,
20 Vauxhall Bridge Road, London SW1V 2SA

Random Century Group Australia (Pty) Ltd, 20 Alfred Street,
Milsons Point, Sydney, New South Wales, 2061, Australia

Random Century Group New Zealand Ltd, 9-11 Rothwell Avenue,
Albany, Auckland 10, New Zealand

Random Century Group South Africa (Pty) Ltd, PO Box 337,
Bergvlei, 2021 South Africa

Printed and bound in Great Britain by Butler and Tanner Ltd, Frome and London

British Library Cataloguing in Publication Data
Wilson, Derek, 1935 -
Breakthrough: Tunnelling the Channel.
1. English Channel. Proposed underwater tunnels. Channel
Tunnel
1. Title
385.312
ISBN 0-7126-3983-7
ISBN 0-7126-3984-5 Pbk

CONTENTS

Breakthrough at Holywell Coombe of the UK north landward running tunnel TBM, signifying the completion of its 8.1-kilometre drive from the tunnelling site at Shakespeare Cliff

The English Channel has formed an often hostile barrier between the British Isles and the Continent for over 12,000 years

CHAPTER I

CHALLENGE

AN ENTERPRISE IN THE NATURE OF THINGS

There is something about the Channel Tunnel that gets into people's blood. For many of the 9,600 construction workers and 3,900 office staff based in Paris and London, Calais and Folkestone, involvement with this project is not just another job. The verdict, echoed at every level in the complex Anglo-French organizations necessary to complete the Tunnel, is that this is one of the great construction projects of all time. The men and women who are making it happen are entitled to feel a sense of destiny.

What, then, makes the Tunnel so special? Why has it gripped the imagination of so many people? The Tunnel will be big - the world's longest undersea tunnel. It will be historic - it breaks a barrier which has stood for well over 12,000 years between Britain and the continent of Europe. It will be timely - it coincides with the Single European Market. It will be a remarkable achievement of international co-operation - French and British participants come to the project with very different attitudes. It will be a technological triumph, one of the most complex private enterprise ventures of modern times. It is, as Eurotunnel's Chairman André Bénard points out, a significant way to end a turbulent century, part of a new European infrastructure symbolizing a degree of unity and common purpose which would have seemed impossible twenty years ago. It should help to change the commercial future of Europe, and should turn the hinterland of the two terminals into rapidly expanding enterprise zones.

Perhaps most remarkable of all, it is a modern success story. For more than 250 years

The UK tunnelling site at Shakespeare Cliff, scene of 24-hours-a-day activity

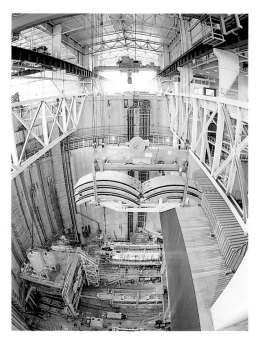

The massive access shaft, 55 metres in diameter, 75 metres deep, at Sangatte, France. This had to be constructed before work on the Tunnel itself could begin.

Views in the early 1880s of the French and British tunnel sites at Sangatte (above) and Shakespeare Cliff (right)

visionaries have dreamed of a fixed link. Many technical, political, financial and communications problems have had to be faced to make it a reality. But now breakthrough has been achieved, Britain is joined to continental Europe for the first time since the Ice Age.

A FALSE START

It all looked very different in 1975: at one end, a roadway, leading to a rectangular opening, closed by solid, spike-topped iron gates across which someone had scrawled the legend NO CHANNEL TUNNEL. That and a few hundred metres of tunnel bored through the chalk on each side of the water was all that remained of almost twenty years of planning, research, fund-raising, political negotiation and preliminary construction.

In the era of national rebuilding which followed the Second World War, the possibility of creating a fixed link across the Channel was seriously considered, though not for the first time. Its champions were reviving the pioneering schemes of the nineteenth century, as we shall see.

Now the time seemed to be ripe. In 1957 two companies, one French, one British, with the approval of the governments in Paris and Westminster joined other powerful commercial interests to form a research group. Some of the world's top experts were involved in a long and careful feasibility study. The result was a decision in July 1966 by the then French Prime Minister, Georges Pompidou, and his British counterpart, Harold Wilson, to proceed with the building of a fixed link. They even announced an opening date - 1975. There was a throb of public interest. Motorists, including several celebrities, booked their places on the first shuttle. One eccentric promoter insisted that he should be at the front of the first train; in the event of his dying before the completion of the project, his coffin was to be dug up and placed in the guard's van!

Yet the time was *not* ripe. Britain was making what appeared to be rather half-hearted attempts to join the European Economic Community (EEC). President de Gaulle vetoed the UK's entry twice. Another Anglo-French prestige project did little to encourage the investment of public funds: Concorde was like a bottomless pit, swallowing up ever larger amounts of development finance. And in 1973 the oil-exporting countries quadrupled the price of oil and sparked off a major economic crisis in the Western world. Nevertheless, Britain *was* now part of the Common Market; in France, Georges Pompidou had become President and, in November 1973, France and Britain signed a treaty for the construction of the Channel Tunnel. But politics is a roundabout. Within months Pompidou was dead and Heath had been ousted by Harold Wilson. The government

lacked enthusiasm for the project and was uncertain of obtaining a parliamentary majority for the necessary legislation. It also wanted to escape from all costly commitments, including the rail link between London and the Tunnel, which it was committed to financing. The treaty was not ratified and in January 1975 work on the project came to a halt.

The French were furious with 'perfidious Albion'. The Paris press accused Britain of wanting to stay aloof from Europe. One French newspaper, *Le Monde*, however, was more philosophical: 'The day will come', it stated, 'when this enterprise, which is in the nature of things and geography, will begin again. But when?'

ENTRENCHED OPPOSITION

If there was dismay in France, there was jubilation among many people in Britain. Barbara Castle, a member of the government, wrote in her diary: 'I am convinced that the building of a

tunnel would do something profound to the national attitude - and certainly not for the better.'

There was understandable opposition from various quarters in the UK - ferry companies, the seamen's union, environmentalists and many inhabitants of south-east Kent - but more than that lay behind the 1975 rejection. It can be summed up in the words of Prime Minister Palmerston who dismissed a much earlier tunnel scheme in 1858 as a project to shorten 'a distance we find already too short'.

France and Britain have not been at war (apart from minor colonial skirmishes) since 1815 and have, in fact, been allies since then in three major conflicts. Nevertheless, old rivalries and animosities were entrenched on both sides of the water. The British called the French 'frogs'. The French dubbed syphilis 'the English pox'. Our mutual history has contained many incidents which have complicated the way we regard each

other: the Norman Conquest; the burning of Joan of Arc; the refuge given to French Huguenots in the seventeenth century; Trafalgar; the D-Day landings; de Gaulle's opposition to Britain's EEC application - to name but a few.

Of course, there is much on the credit side, too: French admiration of the British royal family; English appreciation of French food and wine; friendly rivalry on the rugby field and the thriving tourist trade between the two nations.

Neighbours inevitably squabble and are very conscious of each other's good and bad points. Campaigners for the fixed link have always had to contend with this unique love-hate relationship.

A FRESH START

After 1975 the Labour government had no interest in a fixed link and their French counterparts had no intention of allowing their fingers to be burned a second time. But *Le Monde* was right. This was a project 'in the nature of things'. Other parties were also convinced of its importance and its inevitability.

Among these was the British business community. As trade within the European Community increased, it found itself falling further and further behind its Continental competitors. A major reason was the cost of transporting goods across the Channel. It was not just a question of ferry charges but also the time factor. A lorry could be driven from the Midlands to Dover, Folkestone or Newhaven in four or five hours. Once there, however, it might be delayed as long or longer before being able to proceed those few exasperating kilometres between Britain and France. In fact, the ferry crossing was no quicker in 1975 than it had been in 1875. Almost as frustrating were the problems of business travellers. The air route between Paris and London is the busiest in Europe. Travelling between the two capitals consumed half a working day, involved tedious travel from the centre to the airport, waiting around in terminal buildings and, at the far end, another drive into the city. How much time - and therefore money - might be saved if the journey could be done, city centre to city centre,

in a few hours, in a comfortable train on which one could work, make phone calls and even conduct conferences?

The railway companies also saw the potential which a fixed link would provide for increased passenger and goods traffic. In 1979 British Rail (BR) and the Société Nationale des Chemins de Fer Français (SNCF) produced a joint proposal for a rail tunnel which, because of its narrow diameter, was dubbed the 'mousehole'. SNCF was in the midst of a major modernization programme, the pride of which was its TGV (*Train à Grande Vitesse*), a high-speed train that could run at speeds of 260 km/h (later increased to 300 km/h), and the possibility of a direct Paris-London route was very attractive.

In the early 1980s several committees and study groups from both the banking and parliamentary worlds produced reports on the feasibility of constructing a fixed link, and examined its commercial implications. Several of the proposals came from major British civil-engineering companies. This was hardly surprising since, apart from their genuine expertise, it was also in their interests to promote large construction projects. In contrast

with France, where the early 1980s saw the start of a number of massive state-financed projects, the supply of such work in Britain was dwindling.

Although these adventurous plans all took advantage of the latest developments in technology and construction techniques, none of the basic ideas was new. Perhaps that is not surprising: there is a limited number of ways to convey traffic across a 34-kilometre land gap. You can go over, under or through the sea. You can go by road or rail. These are the options, and they had all been considered before during the previous two and a half centuries.

All this planning and discussion helped to create a favourable climate for the Tunnel. But without some indication of government support, further progress was impossible. When that indication came, it was expressed in words far more positive than the Tunnel's strongest supporters could possibly have hoped for. Meeting at Avignon for one of their regular economic summits, the political leaders of France and Britain issued a statement that 'a fixed cross-Channel link would be in the mutual interests of both countries'.

The formality of these words hid real enthusiasm on the part of both leaders. Mrs Thatcher described the Tunnel as 'a project which can show visibly how the technology of this age has moved to link the Continent and Britain closer together'. For her the Tunnel would boost the private commercial sector and provide tangible evidence of her commitment to the ideal of national recovery spearheaded by free enterprise; from the outset she stipulated that no government money would be available. It would also prove her commitment to Europe in the face of Britain's prolonged political wrangling with her Community partners. In France, the Tunnel was one of several major infrastructure developments initiated in the early years of President Mitterrand's first term in office. The President was following several of his predecessors in giving his personal stamp of approval to a project designed to create nationwide economic and technological benefits with, however, the major difference that, in this case, the state would not have any financial

A 1980s humorous illustration showing work on the Tunnel

Prime Minister Thatcher and President Mitterrand at a ceremony held on 29 July 1987 to mark the ratification of the Treaty that allowed the construction of the Tunnel

The French Tunnel construction site, located next to the small village of Sangatte, near Calais

involvement. In addition, the Tunnel would also help to revitalize the run-down economy of north-west France which was experiencing high unemployment and was not sharing the fast-growing economic prosperity enjoyed by the rest of France. And, for both leaders, the Tunnel also offered the opportunity to leave a permanent legacy of their time in office.

COMPETITION

The two governments decided to stage a competition for the best proposal, to be adjudicated by a joint Anglo-French committee. Now the civil engineering companies went in search of finance houses and banks to support their proposals, and looked as well for partners on the opposite side of the Channel. Amidst frantic rivalry, millions were spent on research, design and lobbying. Everyone was taking an enormous risk. Only one scheme could win, yet each had to be worked out in great detail and, since Anglo-French co-operation was one of the conditions of the contest, constant liaison between Paris and London was essential. It was far from easy to think of everything - and reach agreement. One chairman had to fight hard over a practical matter of facilities. His colleagues suggested that, because the transit would take only about thirty minutes, these should be kept to a minimum. This brought forth a terse memo: 'There must be toilets'.

By the closing date - 31 October 1985 - nine schemes had been submitted. Four were considered worthy of close scrutiny. These were: a motorway suspension bridge; a tunnel accommodating a road and a railway; a combined bridge and submerged tube system; and a tunnel to carry through trains and shuttles for road vehicles.

Between 31 October 1985 and 20 January 1986, when the joint government decision was to be announced, much lavish hospitality flowed in the committee rooms and hotels of Paris and London as competitors with a great deal at stake urgently set forth the merits of their own schemes. The smart money was on the bridge/tunnel scheme (Euroroute), because Mrs Thatcher was known to favour a project which could not be threatened by rail unions and because most motorists would probably prefer to drive straight across. At the last moment, the Channel Expressway proposal (initially for tunnels used by both vehicles and trains) seemed to be making the running. In the end, the winner was the train/shuttle tunnel devised by the consortium Channel Tunnel Group - France-Manche SA (CTG-FM). This proposal later came to be dubbed 'Eurotunnel'.

Shortly after the announcement, the two governments gave their reasons for selecting the CTG-FM scheme. They centred largely on the financing proposals, which were considered to be the soundest put forward, and the practicality of the scheme. It carried the fewest technical risks and was the safest from the traveller's point of view. Unlike a bridge, it would not create any maritime problems, and it was considered the least vulnerable to sabotage and terrorist action.

The three competing shortlisted schemes: Euroroute, a drive-through bridge and tunnel crossing with a separate rail tunnel (top left); Channel Expressway, a drive-through road/rail tunnel (top right) and (below) Eurobridge, a drive-through bridge and separate rail tunnel.

WHAT IS EUROTUNNEL?

The successful scheme, scheduled for completion in June 1993, is linked into the road and rail systems of both Britain and France. It will consist of a central service tunnel and two running tunnels (one in each direction) through which both shuttle trains carrying road vehicles and BR/SNCF through trains will pass.

The shuttle trains will operate in a continuous loop between the terminals at Folkestone and Coquelles, outside Calais. They will offer a frequent service for passenger and goods vehicles, with a planned maximum of seven journeys per hour in each direction at peak times. The road travellers of the future - the very near future - will arrive at the terminal, buy a ticket (there will be no need to reserve in advance), go through frontier controls without leaving their vehicle, and drive on to the next shuttle. At the other end of the 35-minute shuttle journey, they will unload and drive away on to the adjacent motorway.

The through passenger and freight trains between the two countries will be scheduled in between the shuttle services. BR and SNCF have already contracted with Eurotunnel for half the 'paths' through the Tunnel. These trains will all be constructed to the smaller British loading gauge (the maximum width and height of rolling stock); the track gauge is the same in the two countries.

At the start of the 1990s, Europe is experiencing a period of rapid economic, commercial and political change. The countries of Eastern Europe are changing their economic systems. In Western Europe the Single Market will stimulate expansion and change, and industrial and service companies will open new offices and plant on each side of the water. As Europeans take more holidays, the volume of tourist traffic will increase. There simply has to be more mobility. The Tunnel will be a major contribution to a Europe-wide transportation system. It will help tomorrow's Europe to work - and play.

Artist's impression of a passenger shuttle showing double-deck wagons
Dessin illustratif d'une navette passagers montrant des wagons à deux niveaux

1. Locomotive
2. Loading/unloading wagon
   Wagon d'embarquement / de débarquement
3. Door to ramp and upper deck
   Porte d'accès à la rampe et au niveau supérieur
4. Door to lower deck
   Porte d'accès au niveau inférieur
5. Carrier wagon
   Wagon porteur

Photographs of full-sized models of the proposed shuttles: from left to right, a double-deck tourist shuttle; motor cyclists' relaxation lounge; single-deck tourist shuttle.
Photographies des maquettes des navettes proposées; de gauche à droite, une navette de tourisme à deux niveaux; la zone de repos pour motocyclistes; navette de tourisme avec un niveau.

Early nineteenth-century French engraving later used in Britain to capitalize on fears of an invasion through a tunnel

CHAPTER 2

HISTORY

FABLES, FEARS AND FLIGHTS OF FANCY

'Linking France and England will meet one of the present-day needs of civilization.' When the French writer, Louis Figuier, reached that conclusion in 1888 he was only restating a conviction which had been expressed from time to time by many of his compatriots for more than 138 years. It was in 1750 that the Academy of Amiens launched a competition to discover a scheme that would improve trade links between the two nations. The burghers of the city on the Somme, midway between Paris and the main Channel ports, had very good reason to encourage the flow of goods across the Channel. However, when they awarded their prize the following year to the engineer Nicolas Desmarest, can they really have imagined that his proposal was practicable? He was actually suggesting a *tunnel*! The French seldom lack for imagination but this scheme was sheer fantasy. No major tunnelling enterprises had been undertaken since the time of the ancient Egyptians and the Romans - and they had at their disposal large armies of expendable slaves.

The proposal of such a dramatic solution shows just how great the problem was. Britain and France were the world's leading maritime and commercial powers. They were a mere 34 kilometres apart. Yet trade between them was an extremely hazardous affair. The shortest route - across the Pas de Calais, or Straits of Dover - was also the most difficult. Travellers making the current- and storm-beset crossing could, with a fair wind and a skilful captain, be at their destination in six or seven hours. They could equally well be delayed days or weeks - and be exceedingly green by the time they finally set foot on *terra firma*. If only a comfortable and

reliable way could be found of conveying passengers and transporting goods across those tantalizing 34 kilometres without the problems of unloading and warehousing, one of the 'needs of civilization' would, indeed, have been met.

If the advantages were so obvious, why were the British less enthusiastic about a fixed link? Throughout the two and a half centuries of cross-Channel projects it has been the French who have made most of the running. Shakespeare has the answer. In *Richard II* he set before Englishmen the patriotic vision of their land as:

> This precious stone set in the silver sea,
> Which serves it in the office of a wall,
> Or as a moat defensive to a house,
> Against the envy of less happier lands.

The Channel has on many occasions protected the English from their enemies by foiling invasion attempts, most notably the Spanish Armada in 1588 and again in the dark days of 1940 when Hitler's armies overran Northern Europe. But more important than the physical separation of Britain from the Continent was the psychological separation. The English regarded themselves as distinct from and superior to their neighbours 'across the water'. They were semi-detached: part of Europe when it suited them; distinct from Europe when it did not.

This very separateness could also be to Europe's advantage, as was demonstrated during the Napoleonic Wars. As the French emperor carved out a continental empire between 1796 and 1814, Britain alone was able to defy him by controlling the Channel. This provided the basis for the allies' final victory over Bonaparte at Waterloo in 1815. Interestingly, it was during a

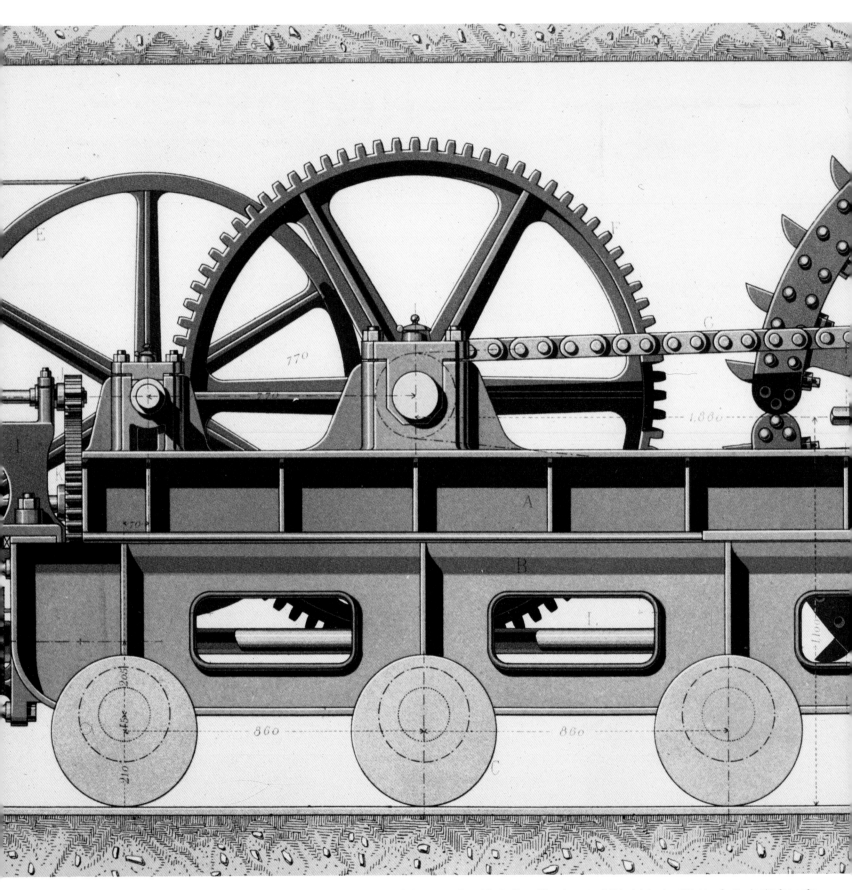

A late nineteenth-century design for a tunnel-boring machine. The 'Beaumont and English' machine which was eventually used for the Channel Tunnel attempt of this period abandoned this type of cutting head in favour of a rotating-disc cutting head, a prototype for all future designs.

break in hostilities that the idea of a tunnel was once again considered seriously. In 1802, Albert Mathieu-Favier, a mining engineer, presented Napoleon with a scheme for two tunnels - one to carry horse-drawn vehicles and, below it, a conduit to carry off water seeping into the tunnel. It had regular air shafts projecting above the waves and provided for an ingenious island town built on a sandbank which would also serve as a free port and artificial haven for shipping. The scheme was supported by the peace party among English politicians. The resumption of the war dashed their hopes. It is doubtful whether many of their countrymen would have supported such a project, especially when they saw French prints depicting the invasion of England by troops carried in ships and balloons *and* marching through a tunnel.

By 1850 Europe was a very different place, confident, industrialized and progressive. The new epoch was one of seemingly unrestricted opportunity for overseas expansion, commerce and industry, and engineering marvels. Railways veined the map of Europe, and with railways came bridges, soaring viaducts and, of course, tunnels. No physical obstacle, whether forest, river or mountain, was allowed to stand in the way of modern communications. When engineers were confronted with new problems they simply invented new machines to overcome them. Nothing seemed impossible.

Yet the Channel still presented the biggest challenge. Numerous bizarre suggestions were made over the years for creating a fixed link. One ingenious engineer proposed laying, in sections, a large iron tube on the sea-bed. Another intended to employ submarines from which men would erect stone piers which would carry a cross-Channel bridge. A variation of the tube idea was to keep the container level by anchoring it to the Channel floor with cables. Yet another device was a huge concrete ferry boat which would carry railway trains. The most straightforward concept of all was that of simply 'filling in' the Channel and making an isthmus pierced by navigation channels. The maritime community was not enthusiastic about this!

Most persistent and dedicated of all these

Albert Mathieu-Favier's tunnel scheme, with candles for lighting and air shafts for ventilation

Illustration of a tunnel proposal made in 1891 in an attempt to overcome military objections. The viaduct at the end of the tunnel could be destroyed in time of war to render the tunnel inoperable.

By the time he presented his proposals to the Emperor Napoleon III in 1867 their feasibility had considerably increased. The great Alpine tunnels of Mont Cenis and St Gotthard were being constructed as were the beginnings of the London underground system. And Ferdinand de Lesseps was busy completing the world's most complex engineering feat to date, the Suez Canal. The demand for coal had resulted in considerable improvements in mining techniques. Indeed, there already existed in one of the Welsh coalfields a tunnel almost as long as the proposed Channel link. No longer was excavation dependent on human labour or the problematical use of explosives; new boring machines had been invented. Long years of peace had improved the political prospects for co-operation between France and Britain. Queen Victoria herself, as someone who suffered badly from seasickness, was a great enthusiast for an alternative means of crossing the Channel. She was also influenced in the middle years of the century by her husband, Prince Albert, who was an eager advocate of progressive schemes. (Prime Minister Palmerston, however, had seen things from a nationalistic perspective and told the consort so. 'You would think very differently if you had been born in this island,' he observed.)

In 1868, Sir John Hawkshaw, one of the finest civil engineers of the day with successful projects all over the world to his credit, set up an Anglo-French group to explore the possibility of a rail tunnel. Having studied previous proposals, the committee came up with a scheme that won the approval of both governments. In 1872 the Channel Tunnel Company was incorporated. French and British railway companies, Rothschild's and other banks, and thousands of private investors subscribed and it seemed that, at long last, the great vision would be realized. Sadly, one who did not benefit was Thomé de Gamond. The man who had done so much to keep the idea of a fixed link alive was ignored by the new generation of engineers and financiers. He fought hard to obtain some recompense for all his work and expense. Then, just at the moment, in 1876, when a French court decided in his favour, he died, worn out by bitter

visionaries was Aimé Thomé de Gamond, a hydrographic engineer. He was obsessed with the problem and for forty years he produced scheme after scheme, but he was far from being a dreamer. He devoted his slender means and considerable technical knowledge to a detailed study of submarine geology. He plunged to the sea-bed himself to collect samples. To stay down as long as possible he filled his mouth with olive oil, the theory being that he could expel air without inhaling water. As a result of his researches, Thomé de Gamond eventually jettisoned all his other ideas in favour of a railway tunnel. The best route, he decided, bearing in mind the geological strata through which the shaft would have to be bored, lay between Cap Gris-Nez and a point on the English coast near Folkestone.

The Brunton drilling machine used in trial bores off the French coast in the 1870s, and later superseded by a rotary machine incorporating features of machines designed independently by Beaumont and English

wrangling.

The work went on. Pilot tunnels were begun. A treaty was drawn up for both governments to sign. But the prospect of a tunnel actually being built, rather than talked about, panicked leaders of British public opinion. Politicians, poets, scientists, churchmen and generals raised loud protest. The old military argument prevailed. A tunnel would bring invasion within the realms of possibility. Britain would have to maintain a large, expensive standing army. In 1882 the government at Westminster put a stop to the digging.

Ironically, it was military lobbyists who campaigned *for* a tunnel during and after the First World War. If Britain had been able to convey men and munitions swiftly to the battle zone, they argued, Germany could have been defeated within months. Winston Churchill, First Lord of the Admiralty, adding his voice in favour, observed that the way to prevent the tunnel being used by an enemy was to have the entrance a quarter of a mile out at sea and linked to the coast by a bridge. Naval gunfire would demolish it at the first sign of trouble. The subsequent development of air power effectively weakened all military counter-arguments. Yet something of the old paranoia remained. In 1941 a member of the War Cabinet insisted that the Nazis were busy tunnelling under the Channel and concealing their activities by washing the chalk spoil away in the sea.

The tantalizing possibility of a fixed link never remained dormant for long. The idea always had its promoters and detractors. But in the inter-war period it was financial objections which held sway. In the Depression years and their aftermath there simply was not the capital available on either side of the Channel for such a vast undertaking. And the old chauvinism was never far below the surface. In 1920 Lord Curzon, a Tory statesman, asserted:
If we could be convinced that we and the French would maintain a perpetual friendship and never quarrel...then it is obvious that there should be a distinct advantage...real friendship has always been very difficult owing to differences of language, mentality and national character....Nor can Great

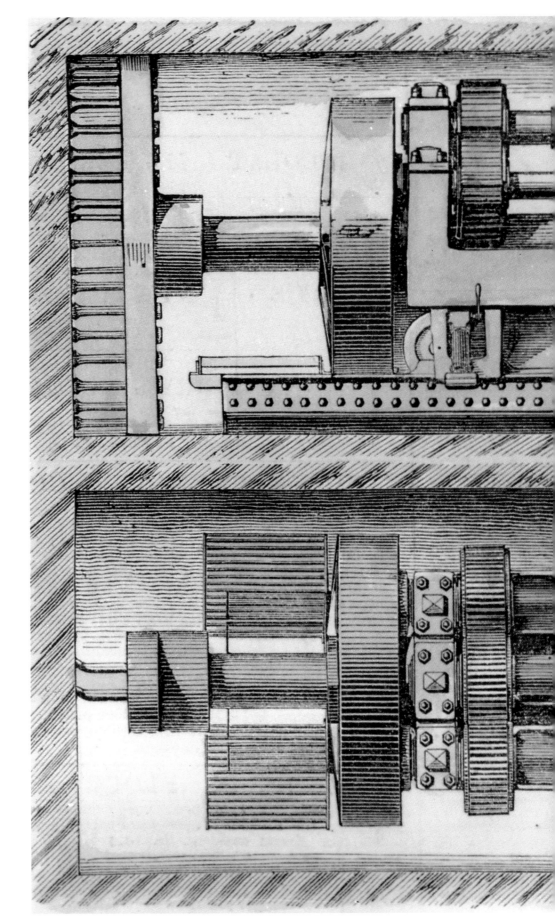

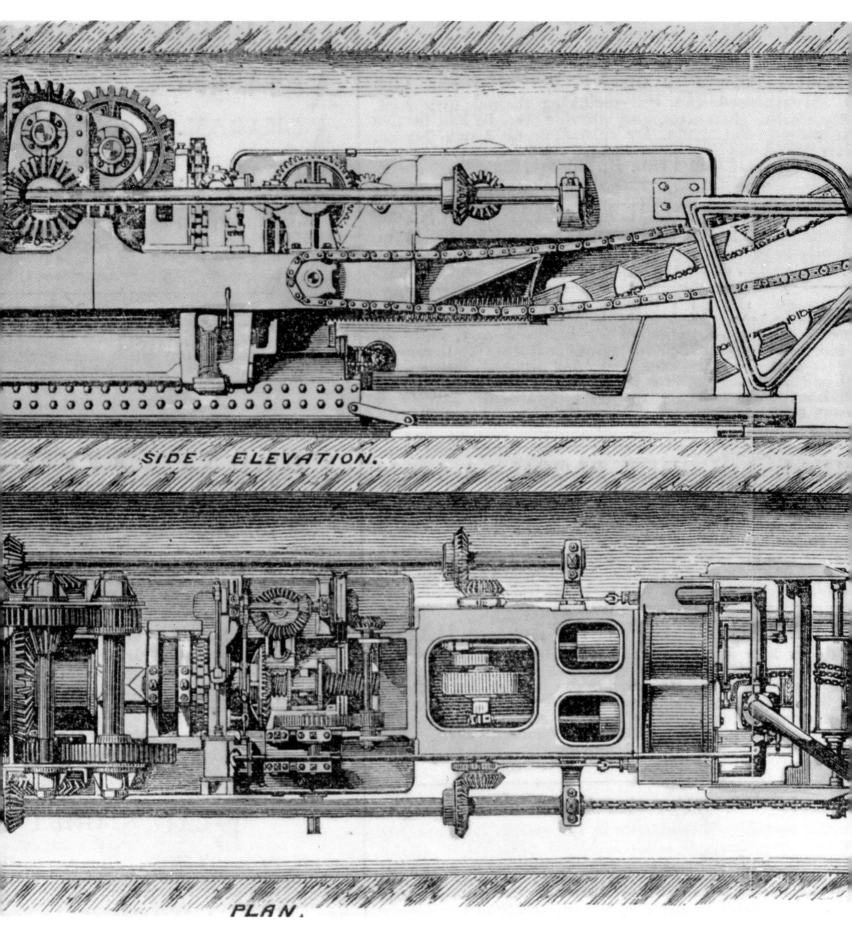

SIDE ELEVATION.

PLAN.

Engraving of the 'Beaumont and English' machine successfully used to construct pilot tunnels in France and the UK in the 1880s

Taking soundings in the Channel in 1875-6 from the Boulogne steamboat Ajax

John Bateman's scheme for a submerged tube anchored to the sea-bed by heavy cables, advocated in the English Mechanic in 1869

Britain place any reliance upon public opinion in France being well-balanced and reasonable.
As long as such attitudes were common there was little prospect of a Channel tunnel ever being built. Only after the Second World War did events bring about a gradual change in British self-perception. The loss of empire, the growth of multinational companies, the importance, first of the North Atlantic Treaty Organization (NATO) and then the EEC, linked Britain's destiny much more firmly to the Continent. By the 1970s not only did the technology for a fixed link exist but the logical inevitability of it had come to outweigh all counter-arguments. As Thomé de Gamond had realized, 'nothing can be done in France for the execution of this project while the English do not take the initiative'. Even that long-suffering visionary would have been astounded to think that it would take a century for that initiative to materialize.

Surveying the geological strata of the sea-bed off Shakespeare Cliff in 1986, from a jack-up rig

CHAPTER 3

THE ENGINEERING TASK

CHALK AND CHALLENGE

LEGISLATION

'The deputies are digging the Channel tunnel' - so ran the headline in *Le Monde* on 24 April 1987. The paper's report told of the almost unique unanimity with which members of the National Assembly had, during the night of 22/23 April, authorized the ratification of a treaty signed by the heads of government fourteen months before. It was in Canterbury on 12 February 1986 that Margaret Thatcher and François Mitterrand signed the Treaty of Canterbury which laid down the legal, financial and administrative basis on which the two nations would co-operate. But legislation still had to pass through the national parliaments before finance could be raised, and it was by no means certain that the representatives of the British and French people would support the fixed link. It was, therefore, in an atmosphere of faith and hope that the originators of the scheme recruited top personnel and started to draw up detailed plans while legislation was prepared.

In fact, government determination and general enthusiasm among parliamentarians were strong enough to see the measures easily through the two parliaments. In both countries efforts were made to expedite the process of obtaining the necessary permissions to proceed with the project. In France two special procedures were used, a *Déclaration d'Utilité Publique* and a *Procédure Grand Chantier*, which enhanced central, as opposed to local, support in the national interest and helped with auxiliary infrastructure, staff training, etc. In the UK all new railway projects normally receive authorization to proceed by means of a Private Bill, a relic of the nineteenth-century railway mania. However, the

Channel Tunnel project involved also an international treaty and frontier matters, all issues that required government intervention. As a result, the necessary legislation went through as a so-called Hybrid Bill, that is, a Public Bill with additional private sections. ('Private' in this context means the promoter of the scheme, in this case Eurotunnel.) As a government bill, it was ensured an all-important place in the parliamentary timetable. An important additional advantage (from the promoters' point of view) was that the consultation and decision-making process was streamlined. The government decided not to hold a public inquiry; instead, objections were considered by select committees of both the House of Commons and the House of Lords as part of the Hybrid Bill procedure. Both committees went to considerable lengths to ensure that objectors received a fair hearing. Nearly 5,000 petitions (that is, objections) were submitted to the Commons Select Committee, and some 1,600 to the Lords. The Commons Committee met on thirty-five different occasions, and held several sessions in Folkestone and Dover.

It is extraordinary that, despite the débâcle of the 1970s, which no one had forgotten, the new fixed-link proposals were carried through with such enthusiasm. French deputies across the political spectrum, from the Communist Party to the National Front, voted in favour and hastened the legislation through. By 6 May 1987 the process in France was complete. In Britain it took a little longer but the Channel Tunnel Act received the Royal Assent on 23 July 1987.

Now, at last, the Treaty of Canterbury could be formally ratified and CTG-FM, which had been

Assembling a UK running tunnel TBM in its underground erection chamber

Lowering part of a French running tunnel TBM down the access shaft at Sangatte

Control console behind the cutting head of a French running tunnel TBM

Right: The cutting head of a French running tunnel TBM ready to be moved along specially laid track for final assembly at the tunnelling face

awarded the concession to construct and operate the tunnel system, could be given the final go-ahead. The system was to comprise two running tunnels, 7.6 metres in diameter, together with a 4.8-metre service tunnel linked by cross passages to the running tunnels at 375-metre intervals. At each terminal the system was to be linked to the national motorway and rail systems. The concession was granted for a period of fifty-five years, in other words until the year 2042.

CORPORATE STRUCTURE

Initially, CTG-FM consisted of a consortium of British and French banks and construction companies, made up as follows:

CHANNEL TUNNEL GROUP:
  Balfour Beatty Construction Limited
  Costain Civil Engineering Limited
  Tarmac Construction Limited
  Taylor Woodrow Construction Limited
  Wimpey Major Projects Limited
  National Westminster Bank plc
  Midland Bank plc
FRANCE-MANCHE SA:
  Bouygues SA
  Dumez SA
  Spie/Batignolles SA
  Société Auxiliare d'Entreprises SAE
  Société Générale d'Entreprises SGE
  Banque Nationale de Paris
  Crédit Lyonnais
  Banque Indosuez

Once the two governments had decided to go ahead with the project, the shareholders of CTG-FM realized that they would have to separate the roles of owner/operator and constructor. This general intention was made more difficult because the original shareholders were either construction companies or banks - neither natural owners nor operators.

Large sums of money were at stake and at times the process was acrimonious. Eurotunnel SA (in France) and Eurotunnel plc (in Britain) were formed as the umbrella holding companies with the shareholding of the contractors and banks in Eurotunnel diminishing as Eurotunnel

Surveying work adjacent to the French Tunnel portal

A diver prepares to examine the piling of the spoil lagoons at the Shakespeare Cliff site

Right: Welding in a French tunnel

transformed into a separate independent company. The contractors in the original CTG-FM group formed themselves into a two-nation engineering and construction contractor, Transmanche-Link (TML). A contract for the design, construction and commissioning of the complete tunnel system was then negotiated between Eurotunnel and TML.

As all this was happening, both Eurotunnel and TML had to put in place appropriate corporate structures staffed by the best experts available. It was a time-consuming and at times difficult process, and not without some casualties. The Eurotunnel drama has been performed on several stages - and only one of them lies beneath the waters of the Channel.

FINANCE

In financial terms as much as in the engineering, the Channel Tunnel is an unprecedented venture: a transport infrastructure project developed and financed by the private sector alone, without any form of financial support from either the French or British government. A financing scheme formed an integral part of the original submission presented by CTG-FM to the French and British governments in October 1985. The scheme provided for the estimated cost of the Tunnel to be financed by £5 billion worth of bank loans, with equity of a further £1 billion coming from the promoters, institutional investors and a public share flotation.

Preliminary equity finance was raised in two main stages, known as Equity 1 and Equity 2, in September and October 1986. Equity 1 was subscribed by the original banks and construction companies involved in the project, while the cash for Equity 2 came principally from French and British investment institutions.

Much of 1987 was taken up with the mammoth task of negotiating a banking credit agreement and organizing Equity 3, Eurotunnel's major equity fund-raising and simultaneous listing on the Paris and London stock exchanges. On the banking side, teams of lawyers, financiers and technical experts from Eurotunnel and the arranging banks and merchant banks worked for months to settle all the highly complex detail and prepare the necessary documentation. The banks agreed to underwrite the credit agreement in August 1987. This paved the way for the syndication of the loan to banks throughout the world. Commercial, financial and technical presentations - known as 'syndication roadshows' - were held around the world. The target of syndicating 50 per cent of the banks' original commitments was achieved for signature in early November with a margin to spare. This increased the total number of banks in the syndicate to 198.

During 1987 parliamentary approval of the Tunnel project had to be obtained in both countries. Stock market conditions were kept under review, and on 19 October - 'Black Monday' as it soon became known - the international share market collapsed and billions were wiped off share values. However, the decision was taken to continue to proceed with Equity 3 on the planned date, 16 November. Postponing the share issue could have been fatal for the project itself, and the investment (with dividends not expected to be paid until some years after the opening of the Tunnel) was longer term than the current market situation.

The Paris stock exchange, La Bourse

In the end, over 300,000 investors (two-thirds of them in France) bought shares, but a portion of the issue was subscribed by the underwriters, who stepped in to take up the balance not applied for by the public. Two crucial events during 1987 helped to boost the project's prospects. The first was the signature of the Railway Usage Contract with British Rail and SNCF, which gave Eurotunnel a guaranteed income from its largest single customer; the railways were expected to provide about 40 per cent of the company's planned revenue and use half the capacity of the Tunnel. The second was the French government's decision to build the TGV nord line from Paris to the Belgian border, with a spur through Lille to the Tunnel mouth, thus enabling the railways to offer a three-hour service between London and Paris: real competition for the airlines.

In July 1989, Eurotunnel announced that additional funding would be necessary as a result of substantial increases in the forecast cost of the Tunnel. The contractors, TML, later agreed to meet a fixed percentage of cost over-runs in tunnelling works, and a range of cost-saving measures was also announced. In June 1990, it was announced that the funds available to Eurotunnel needed to be increased from £6 billion to about £8.5 billion, well above the then forecast cost of £7.66 billion (June 1990) required to complete the project. Detailed negotiations on this additional funding were still proceeding as this book was sent to the printer.

CHOOSING THE ROUTE

It comes as a surprise to learn that of all the challenges presented by this mammoth project the least testing is that of engineering. Solving logistical, financial, environmental and other problems has created considerable headaches, but in terms of a pure tunnelling exercise the fixed link has called for little in the way of innovative technology. That certainly does not mean that what is happening 40 metres below the sea-bed is anything other than extremely impressive to the layman. The massive tunnel-boring machines (TBMs), the laser alignment of the boring process, the highly skilled gangs securing the lining segments in place, the co-ordination of a thousand and one jobs create an atmosphere of purposeful, hi-tech efficiency. But to the experts all this is nothing new. The experience and the equipment have been perfected on other jobs around the world, and the sites themselves are comparatively uncomplicated.

The gleaming 'white cliffs of Dover', which

One of the Eurotunnel travelling exhibitions staged to raise public awareness in 1987

Steering a service tunnel TBM (detail greatly exaggerated)
Pilotage d'un tunnelier creusant la galerie de service (détail amplifié)

Surveying the tunnel (simplified)
Levé de la galerie de service (simplifié)

1. Planned route of tunnel
   Tracé prévu de la galerie de service
2. Machine slightly off its true line
   Tunnelier légèrement déporté de son tracé prévu
3. Surveyor's laser beam
   Rayon laser de l'arpenteur-géomètre
4. Computerised TBM alignment target for laser
   Mire d'alignement informatisée du tunnelier pour le laser

5. New line of tunnel, on course
   Nouveau tracé de la galerie de service à suivre
6. TBM cutting head, rotating at 2-3 rpm
   Tête de coupe du tunnelier, tournant à une vitesse de 2 à 3 tours/minute
7. 8 individually adjustable hydraulic rams each working at up to 3000 psi
   8 vérins hydrauliques individuellement réglables, travaillant chacun à une pression atteignant 210 kg/cm2
8. Telescopic section
   Section télescopique

9. Main TBM body with hydraulic gripper rams and gripper pads
   Corps principal du tunnelier équipé de vérins d'ancrage hydrauliques d'ancrage
10. Tunnel lining segments
    Voussoirs de revêtement de la galerie de service
11. TBM back-up equipment omitted for clarity
    Train technique du tunnelier non-illustré pour rendre le schéma plus c

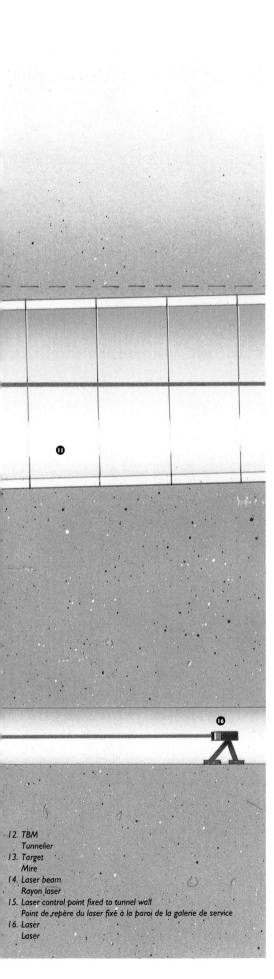

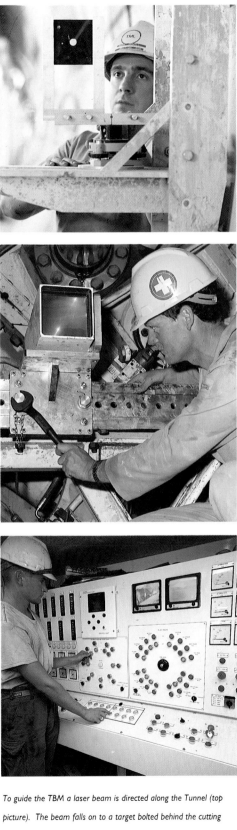

To guide the TBM a laser beam is directed along the Tunnel (top picture). The beam falls on to a target bolted behind the cutting head of the TBM (middle picture), which transmits the data via a computer to the operator's console in the TBM, enabling continuous minute corrections to be made to the Tunnel alignment (bottom picture).

over the centuries have provided foreigners and homecomers with their first glimpse of England, are the visible evidence of part of the chalk strata that form the sea-bed and run almost all the way to the French coast. This 'middle chalk', to give it its geological name, is poor material for boring tunnels through as it is both brittle and porous. Beneath it, however, lies the 'lower chalk', and the base of this stratum, comprising a mixture of chalk and clay called chalk marl, is very different. It is strong, almost impervious, slightly plastic and, therefore, less susceptible to fracture. It would be hard to find a better medium for tunnelling purposes. The marl, in turn, is bedded on gault clay, a weak and unpredictable rock. Thus engineering strategy is clearly dictated by the geological structure: as far as possible, follow the chalk marl.

Some of the very few visitors who receive the necessary security clearance to enter the underground workings are surprised that the three parallel seaward tunnels do not run arrow-straight through the rock, but curve gently up and down and side to side. The reason for this apparent waywardness is, of course, that the engineers have to follow the geology. The chalk marl stratum undulates. So must the tunnels. In this instance the shortest distance between two points is not a straight line.

All this had been known for well over a hundred years. Thomé de Gamond and other pioneers had taken rock samples in the mid-nineteenth century. Colonel Beaumont of the Royal Engineers made drillings in the sea-bed in preparation for the 1870s tunnel scheme. That project had actually pioneered the route from Sangatte to Shakespeare Cliff that CTG-FM chose - using all the best twentieth-century technology. Beaumont drilled over 2,000 metres towards France and it is a tribute to the engineers of the time that much of his tunnel is still intact. Today's technologists were able to secure a fuller understanding of the geological structure. They took core samples from beneath the sea-bed from jack-up platforms, rather like miniature oil rigs. By submitting these specimens to various tests they were able to calculate how the rock itself would react to drilling. Using

The erection chamber of a UK landward running tunnel TBM seen prior to the installation of lifting hoists

seismic equipment, they were able to detect faults and fissures. But, as Colin Kirkland, Eurotunnel's Technical Director, says, this research produced few surprises.

Despite this, nothing could be left to chance. Even with all the probing and testing that had already taken place over many decades, it was essential to ensure that the construction methods could cope with any unexpected geological eventuality, such as a fault in the sea-bed. For this reason, small-diameter probe holes are drilled ahead of the service tunnel boring machines (TBMs) to check that there is no poor ground ahead that could cause a problem for the TBMs. Any fractured areas found by the probes are 'grouted up' with either a mixture of cement and water or with other special chemicals. In addition, the TBMs on the British side have watertight doors which can be closed in the very unlikely event of a major flood risk. The French TBMs are, in any event, designed to operate under the fullest head of water possible. These precautions proved their worth at an early stage at the British side, when the tunnellers on the service tunnel encountered a section of chalk marl that was weak and water-riven. This slowed progress quite considerably for several weeks.

WATER SEEPAGE

Beaumont and his French colleagues discovered that the chalk marl, although continuous across the Channel, dipped near the French coast to be replaced by porous water-bearing strata. In addition, the chalk marl itself was more fractured and faulted nearer the French side. Thus tunnellers from the French side in the 1880s had to drive their first headings under the Channel through ground with water seeping in at the rate of one litre per minute per metre. This contrasts with the British side, where the 1880 tunnel remains intact and dry even to this day. This means that working from each side calls for different techniques. On the French side, the TBMs are designed to bore and line the tunnels with pre-cast concrete or cast-iron linings while operating in totally water-logged ground, even with water under high pressure. On the British

side, with better ground, simpler and faster machines were considered more suitable. These also bored the tunnel and erected similar cast-iron and pre-cast concrete linings. These concrete linings were similar to those used in the recently constructed London tube lines.

SPOIL DISPOSAL

If some problems have become easier to solve with the passage of time and the acquisition of knowledge, others have become more difficult. A hundred years ago Calais, Dover and Folkestone were small fishing ports with a hinterland of sparsely populated farms. No one had ever heard of 'environmental protection'. The process of acquiring land for the latest engineering project was fairly simple. There were few restrictions placed in the way of technological progress. Wealthy industrialists could lay railways across the countryside, pile up slag heaps or abandon worked-out quarries with very few legal restrictions. When the Channel Tunnel Company and the Association du tunnel sous-marin started their abortive project in the 1870s, they adopted the most obvious solution to the spoil-disposal problem; they dumped it in the sea.

Today such an action would be out of the question because of its devastating effect on the marine ecostructure. So an important part of the scheme was the methods proposed for dealing with the spoil coming up from the tunnels. The amount of chalk spoil had to be carefully calculated, and Eurotunnel had to show that it could be disposed of in an acceptable way. At Sangatte it was possible to find a land site close to the construction site. On the English side, however, space is much more restricted, and Eurotunnel came up with an imaginative reclamation scheme which involved extending a new land platform out to sea from the base of Shakespeare Cliff.

THE TERMINALS

It was not only the Tunnel that had to be located as practicably and economically as possible; the terminal areas at each end of the fixed link also had to be sited carefully. Sites suitable for access

to the two national road and rail networks were essential. In addition, there had to be sufficient space for all the necessary buildings and services and, as far as possible, they had to avoid existing centres of population.

Here the French had a distinct advantage. The 1974-5 project had designated an ideal terminal site at Coquelles, some 3 kilometres from Calais. It was set in a broad, level plain of farmland and marshland and the then concessionaires had acquired ample land for the buildings, platforms, bridges and approach roads and rails. On the debit side, this corner of France was inadequately served by roads and railways so a new motorway and high-speed railway line were built into the plans from the beginning.

The British in Kent had to bring the Tunnel under the Downs to Folkestone where a much smaller terminal of approximately 140 hectares could be fitted in between the edge of the town and the hills. Unfortunately, this involved considerable disruption for the adjacent villages - Newington, Peene and Frogholt (see Chapter 6). The site lay adjacent to the existing M20 motorway and the railway line to London. Although there were immediate advantages to this, the road and rail links still needed upgrading to cope with the rapidly increasing volume of local and cross-Channel traffic, including that using the Tunnel. The intrinsic difficulties of the Shakespeare Cliff and Folkestone sites would present the planners and construction teams with many problems. As we shall see, building the fixed link called for a combination of traditional skills and ingenuity.

Geological cross-section of the tunnel route
Coupe géologique - tracé du tunnel

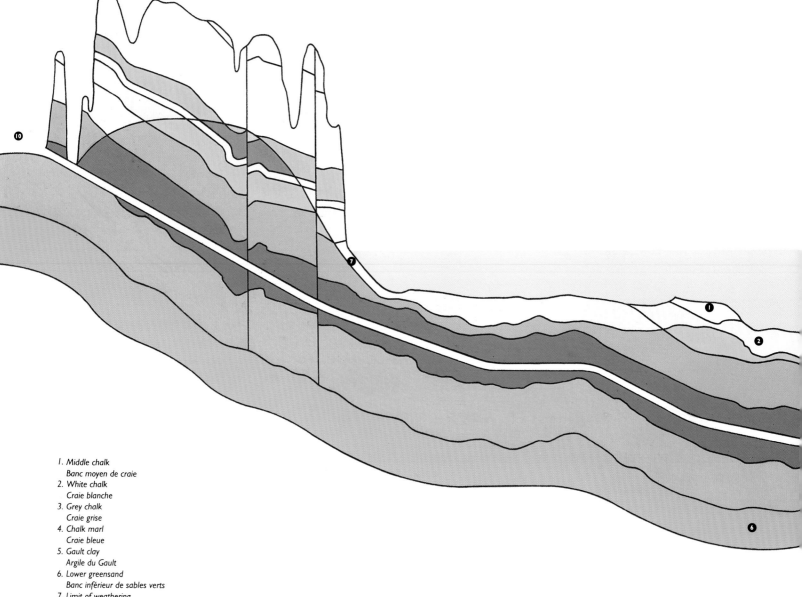

1. Middle chalk
   Banc moyen de craie
2. White chalk
   Craie blanche
3. Grey chalk
   Craie grise
4. Chalk marl
   Craie bleue
5. Gault clay
   Argile du Gault
6. Lower greensand
   Banc inférieur de sables verts
7. Limit of weathering
   Limite des phénomènes d'altération
8. Sea level
   Niveau de la mer
9. Alluvium
   Alluvions
10. Folkestone terminal
    Terminal de Folkestone
11. Coquelles terminal
    Terminal de Coquelles

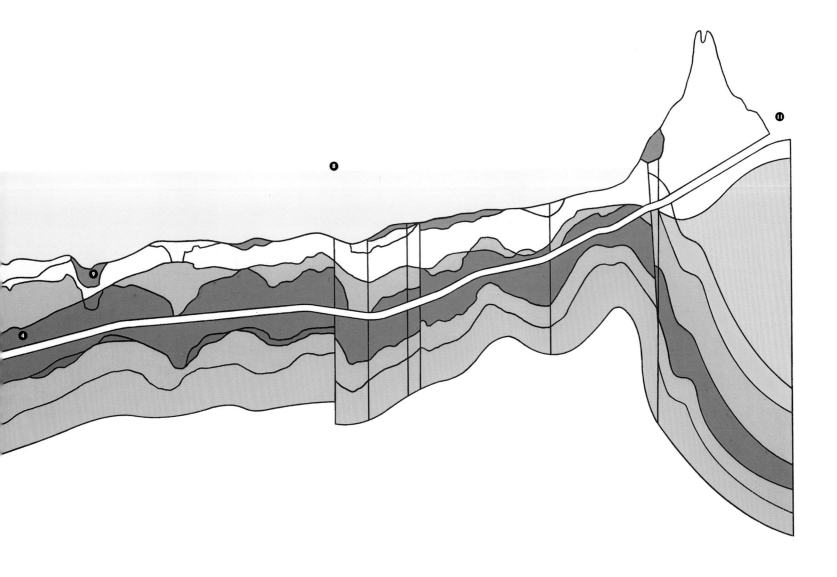

A UK seaward drive running tunnel TBM
Le tunnelier conçu pour creuser l'un des deux tunnels ferroviaires sous mer, côté britannique

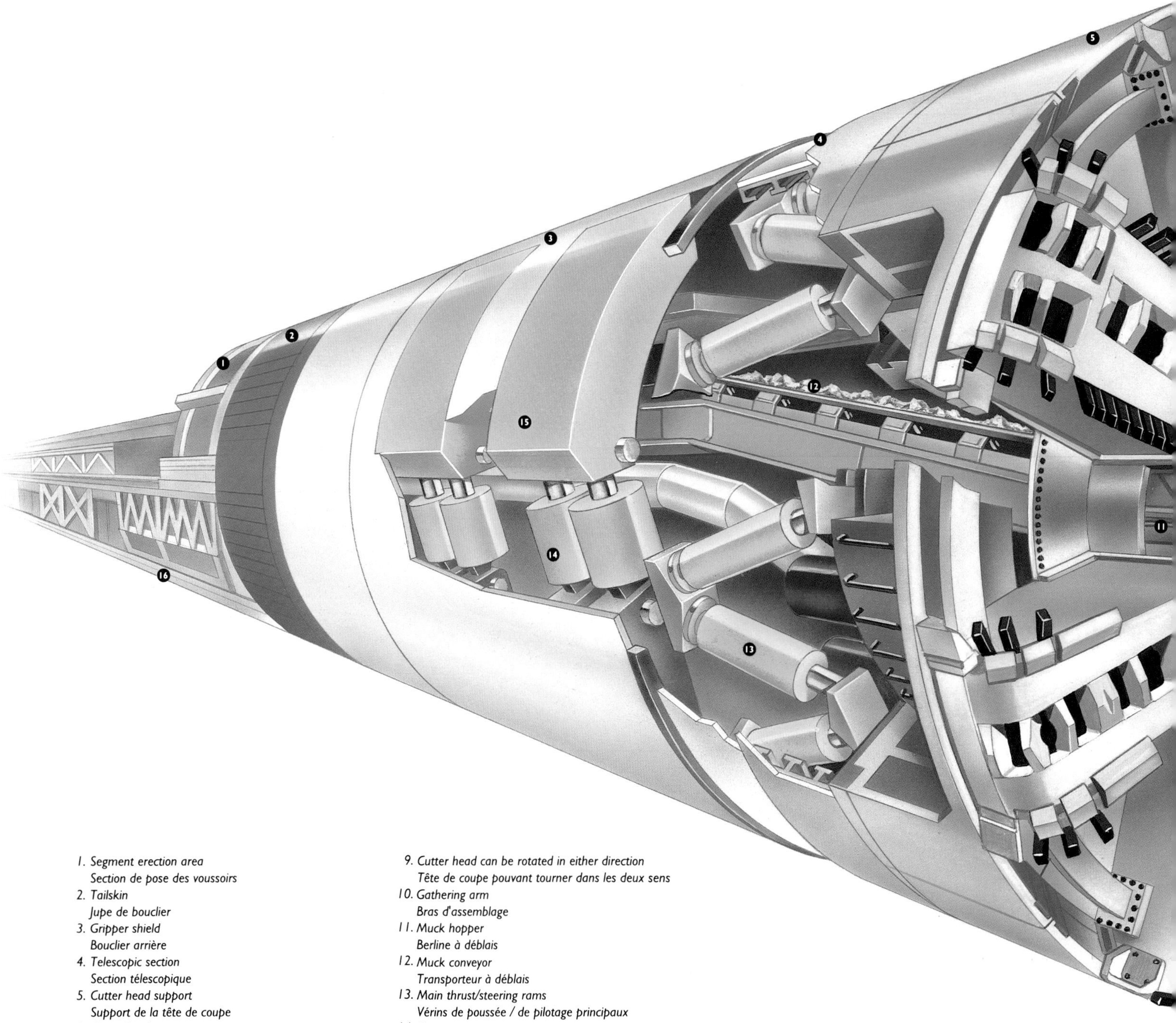

1. Segment erection area
   Section de pose des voussoirs
2. Tailskin
   Jupe de bouclier
3. Gripper shield
   Bouclier arrière
4. Telescopic section
   Section télescopique
5. Cutter head support
   Support de la tête de coupe
6. Cutter head
   Tête de coupe
7. Cutting teeth
   Dents de coupe
8. Cutting scraper
   Raclette de coupe

9. Cutter head can be rotated in either direction
   Tête de coupe pouvant tourner dans les deux sens
10. Gathering arm
    Bras d'assemblage
11. Muck hopper
    Berline à déblais
12. Muck conveyor
    Transporteur à déblais
13. Main thrust/steering rams
    Vérins de poussée / de pilotage principaux
14. Gripper ram
    Vérin d'ancrage
15. Gripper pad
    Patin d'ancrage
16. Back-up structure and services
    Train technique et équipements divers

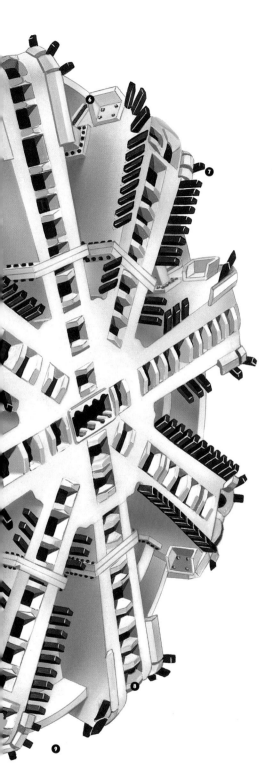

CHAPTER 4

MEN AND MACHINES

A MOLE CALLED BRIGITTE

'This is actually an easy tunnel. It's just a long, long way.' Those words of Peter Bermingham, Superintendent of Land Tunnels at Folkestone, put the fixed link very neatly into perspective. The tunnelling work is not difficult for experienced tunnellers. This is not even the world's longest rail tunnel; that distinction belongs to the Seikan Tunnel which links the Japanese islands of Honshu and Hokkaido. But it *is* the longest piece of underground boring ever achieved without intermittent shafts. This is what, in professional terms, makes the Channel Tunnel special. It creates a staggering logistical problem. Most tunnels have their services (materials, ventilation, power cables, equipment, etc.) fed to them by means of vertical shafts spaced out along their route. Everything the fixed-link tunnellers need has to go underground at Sangatte or Shakespeare Cliff. It has to supply a workforce getting further and further away every day. The supply line is eventually extended over 20 kilometres.

Nor is it just a question of getting everything down to (and up from) the Tunnel; it all has to be shifted at the right time, at the right speed and in the right sequence. Any delay has a knock-on effect and can very quickly bring work to a halt. John Gleeson, a voluble Irishman and General Foreman at Shakespeare Cliff surface site, knows all about that. It is his job, throughout a twelve-hour shift, to control the flow of services by train down the English end of the Tunnel.

As soon as you start work you're totally switched on. You've got to be. Everything's got to go down as and when required. That's the deal. If you're ten minutes late with something that's serious. Then the next train is ten minutes late, and the train movements get fouled up. There's limited track access to get the gear down. You really have to get it right. This kind of work is bastard work! It is! But that's what drives you on. It is different. You're not in a place where you can say, 'Well if it don't get done today, it'll get done tomorrow'. This kind of work is total, bang, bang, bang, bang, up down, up down. And that's the challenge of it.

MODERN TECHNOLOGY

Another thing that makes this job different is the sophistication of the equipment and technology being used. To fix the position of the Tunnel the surveyors used satellites. During 1986 and 1987 several observations were made with the aid of the Global Positioning System. This made possible a greater degree of accuracy in the horizontal control network of the Tunnel. In fact, the degree of error was reduced to about one part in a million.

It is one thing to line up the beginning and end of the Tunnel on the surface, but how is that done underground? The most commonly asked question is, 'How do you get two tunnels to meet in the middle?' and many are the jokes and cartoons showing French and British tunnellers missing each other. One of the earliest such stories was of the engineering company who tendered for the job and submitted a very low price. When asked if they had planned sufficiently carefully for linking their converging tunnels, they replied nonchalantly, 'Oh well, if we miss you'll get two tunnels for the price of one'. Another question that sometimes occurs to the layman is, 'Why meet in the middle, anyway?

Arrival of a section of one of the two seaward drive running tunnel TBMs at the Sangatte site

Wouldn't it be simpler to start at one end and dig through to the other?' The answer is that it would be much more difficult. We come back to logistics again. It is a big enough problem getting men, machines, material and equipment along a tunnel that could, eventually, be up to about 20 kilometres long. Double the length of the tunnel and you more than double the problem and the programme period. A two-ended tunnelling operation is faster, more efficient and, therefore, more economical.

An intermix of modern technologies is used to keep the tunnels on their proper course. The exact route has been planned in detail before the tunnel is begun. Then, as the tunnel-boring machine (TBM) moves forward, the rockface ahead is probed up to a distance of 150 metres with narrow drills which sample the chalk marl and indicate any patches that should be avoided. As the machine grinds forward there is never less than 20 metres of probed ground between it and the virgin rock.

As the TBM pursues its predetermined, gently undulating course it is constantly being aligned from behind using laser technology. A beam is directed at a light-sensitive target on the TBM. As each metre is completed a computer aboard the machine calculates the direction to be drilled in accordance with its predetermined data. Simultaneously, the precise position as revealed by the laser beam is logged. The two sets of information are compared and indicated on a display in the driver's cab. Some minute adjustments and the blind mole continues on its predestined path. This combination of modern technologies enables a tunnel to be cut for 22 kilometres and to meet another such tunnel coming the other way with only a few centimetres of error in alignment.

TUNNELLING STRATEGY

The fact that there can be no service shafts along the length of the undersea tunnel means that access to the underground workings has to be as close as possible to the shorelines of Nord-Pas-de-Calais and Kent in order to minimize the logistical problems as far as possible. Access shafts are located at Shakespeare Cliff and at

Foot of the access shaft at the Sangatte site, looking towards the three landward tunnels

Sangatte. From these points boring takes place in two directions: marine tunnels are drilled under the Channel and, simultaneously, landward tunnels are cut towards the terminal sites at Folkestone and Coquelles.

The system consists of three tunnels. The two outer tunnels, which will carry the shuttles and trains, are known as running tunnel north and running tunnel south. They have a lined diameter of 7.6 metres, and are each separated by 8 metres of rock from the smaller central service tunnel (lined diameter 4.8 metres), which will be used for maintenance and emergency purposes. It is this service tunnel which was cut first and which remained ahead of the running tunnels. It was, if you like, rather like the tip of a triangular arrowhead with the running tunnels being the outer points. The service tunnel acted as a pilot tunnel, testing out the geology before the big running tunnel TBMs arrived.

Crossovers at each end of the Tunnel, and also at two points along its length, enable trains to switch from one track to another. This is so that single-line working can be maintained if a section of tunnel has to be closed for maintenance or in the event of an emergency. The two undersea crossovers are immensely impressive cathedral-like caverns, which the traveller will hardly see, and constitute perhaps the greatest single engineering accomplishment in the whole project. Moreover, they had to be constructed in a way that caused as little disruption as possible to the progress on the tunnels. What that means can be seen by some sample figures: in one month (March to April 1990) 6,000 cubic metres of chalk had to be removed from the British crossover and 1,000 cubic metres of concrete taken in.

CROSS TUNNELS

As well as the main tunnels there are two types of linking tunnels. At intervals of 375 metres there are cross passages which join the running tunnels to the service tunnel. These cross passages will provide easy access for maintenance and will be a means of evacuating passengers in an emergency. Every 250 metres, piston relief ducts arching over the service

Partial assembly at Dunkirk of a French seaward drive running tunnel TBM, following its arrival by sea from Japan

Lowering part of a TBM down the 110-metre access shaft at Shakespeare Cliff (above); hoisting the cutting head of a TBM into position prior to assembly at Shakespeare Cliff (below)

tunnel connect the railway tunnels. These are to reduce the build-up of air pressure as trains and shuttles hurtle through the tunnels at speed.

These smaller diameter tunnels require different construction techniques. They are cut using hand tools and smaller machines. It is here that skills acquired in mining come into their own. Sophisticated technology and traditional methods rub shoulders under the Channel.

THE TBMS

In terms of equipment, the most impressive items are, by a long way, the massive tunnel-boring machines. The biggest is 8.72 metres high and, including its service train, some 260 metres long, which is rather more than two football pitches. The tungsten-carbide teeth of the cutting head are capable of biting their way through the chalk marl at more than 1,000 metres a month. Each of these gargantuan moles, packed with sensitive and sophisticated equipment, is guided by a driver using fingertip controls and aided by computer

displays and television monitors which, among other data, show the cutting-head performance characteristics, movement, setting out, spoil disposal and details of the environment in the tunnel. Each of these machines is individually designed for the job it has to do; the total cost of a UK running tunnel TBM is about £8.5 million. The TBMs for the marine running tunnels are considerably larger than the ones used for the service tunnel and are different again from the landward running tunnel TBMs, which are even larger. Furthermore, there is a very significant distinction between the British machines, which can operate in the open mode, and those employed in the French marine tunnels, which can operate in the closed mode. The basic technology is British and six of the eleven TBMs come from factories in England and Scotland, but it is the Japanese who have perfected the means of coping with the particular problems of boring through the very porous water-bearing rock on the French side. Let us take 'Brigitte', for example. The French have given names to all

Unloading a TBM section at Dover docks, for transporting by road to the Shakespeare Cliff site

Rear view of assembly of the UK north seaward running tunnel TBM (left); part of the French north seaward running tunnel TBM at the base of the shaft, prior to assembly in the tunnel (top left); the cutting head of the UK north seaward running tunnel TBM nearing completion (top right)

their TBMs (Brigitte, Europa, Cathérine, Virginie, Pascaline). 'Brigitte' was the pioneer heroine which carved out the service tunnel and shared in the breakthrough triumph. 'Brigitte' was designed to operate in either of the two modes. As she advanced through the water-bearing and fissured chalk she worked in closed mode. That means that both the cutting head and the body of the machine were sealed so that they were resistant to water, which otherwise would have entered the TBM (and the tunnel) at a pressure of up to 11 kilograms per square centimetre. The entire process from boring to lining the tunnel with concrete segments was performed from within this waterproof module. The spoil (that is, the cut chalk marl) was mixed at the cutting face with water under high pressure to form a liquid mud which was then pumped into trucks which conveyed it back along the tunnel. Once 'Brigitte' had safely reached the chalk marl stratum she was able to revert to open mode. Like the British TBMs she could operate in a more straightforward way, without seals, and

evacuate spoil in solid form. Working in closed mode was a slower process than working in open mode (the rate of progress was almost halved). That is why it was always projected that breakthrough would take place closer to France and not in the middle of the Channel.

Throughout the period of operation these huge machines are kept working virtually round the clock by teams of shift workers. For example, 'Europa', which is boring one of the French marine running tunnels, has two crews of fifty men each, who keep the TBM going for 17 hours in every 24. The remaining period is devoted to maintenance. Various components need replacing on a regular basis if breakdowns are to be avoided. By the time 'Europa' completes her task very few of her parts will be the ones she started out with. The continuous rebuilding is, in itself, something new. On most tunnelling jobs TBMs are rarely asked to travel more than 7 or 8 kilometres. They can then be overhauled, usually in specialist workshops. But these TBMs must cover up to three times this

distance. For them there is no escape from the tunnel, no respite.

To some extent the same is true for the crews, the maintenance engineers and the supply teams who must keep 'Brigitte' and her sisters supplied with all their needs along lines of communication becoming hourly more attenuated. You do not have to talk to tunnellers for very long to realize that they are a different breed of men. Among construction workers they are unique. Highly paid? Certainly. A British TBM crew member can earn up to £1,200 a week. But for that money he spends two-thirds of every working day underground, is probably separated from his family for weeks or months on end, has only his workmates for companions on and off duty, lives the artificial life of the construction camp, works extra hours when the job demands and is constantly under pressure.

Yet, strangely, the pressure is one of the things experienced tunnellers love. They are extraordinarily positive men. They have a tremendous sense of pride in their achievement,

Part of the support train of the French south seaward running tunnel TBM at the base of the access shaft. The support train is towed behind the TBM cutting head and contains equipment to supply and service the Tunnel construction.

and a continuous competitive urge to achieve more than the next crew. John Gleeson explains tunnelling as

unlike any other civil-engineering work. You're always flying by the seat of your pants and the adrenalin flows quite a lot. There's never a dull moment. The one different thing about tunnel work and the people in it is that they want to get the job done and they want to get it done now. When you change your gear in the morning you want to do x amount of work - plus. You don't just get as much work done every day as you can. You don't settle for a certain amount. Tunnellers are people who want jobs done. They don't want to be associated with a job that doesn't get done. They want to do more work today than they did yesterday.

This urgency is traditional in the industry because every company that commissions a tunnel wants it completed in the shortest possible time. No revenue starts flowing in until the tunnel is finished, especially when it is a vital part of a transport network which cannot come into operation until every segment is in place.

The Channel Tunnel is no exception. Urgency is communicated to the contractors. The contractors communicate it to local management. Management communicate it to the on-site workforce. Urgency leads to competition between teams and between the British and French workers. At Sangatte and at Shakespeare Cliff the progress of every shift, at both ends of the Tunnel, is prominently displayed. That is why, throughout the project, performance has constantly been improved and records have regularly been broken. Windsor Jenkins, the TBM electrical manager at Shakespeare Cliff, recalls that, when he worked on the Dartford Tunnel under the Thames fifteen years ago, 1 metre of tunnelling (for their size of tunnel) per shift was considered good progress. Workers on the fixed link would not be satisfied with less than 20 metres a shift. In April 1989, the TBM 'Virginie' made the first breakthrough on the French side when she breached the chalk wall at Coquelles, thus completing the landward service tunnel, having already achieved the fastest ever monthly drive of 886 metres. That record was soon looking

Reassembly at the French portal of the landward running tunnel TBM. Since the French landward tunnels are only about 3 kilometres long, the construction schedule allowed this TBM to be used twice, to bore both French landward running tunnels. Therefore, unlike all the other TBMs, this one will finish where it started.

Dismantling the French landward service tunnel TBM, the first to complete its drive(above); removing the cutting head of the French landward service tunnel TBM from the portal (below)

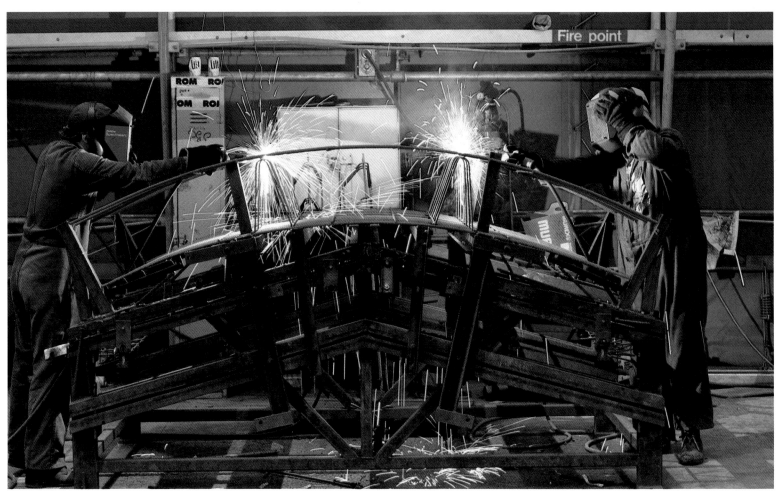

Making wire reinforcement cages for the UK concrete tunnel-lining segments at the Isle of Grain plant

Concrete tunnel lining segments at the storage area behind the production sheds at the Isle of Grain plant, awaiting delivery by rail to the Shakespeare Cliff site

Segment-handling facility at the segment storage area at Sangatte. Unlike its UK counterpart, the French concrete segment plant is adjacent to the tunnel access shaft.

A train of segments and other materials en route to the TBM. These trains are hauled by diesel or electric locomotives running along temporary tracks.

Placing a concrete tunnel-lining segment

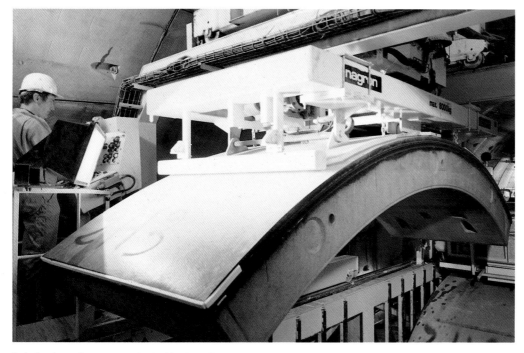

In the French tunnels suction is employed to lift and transfer the segments from the supply train forward to the area behind the cutter head where the tunnel lining rings are erected

Preparing for the injection of cement grout behind the segment ring in a running tunnel

Inserting a smaller 'key' segment in order to expand and lock the tunnel lining ring in place

very insecure. In June 1990, tunnellers on one of the British land tunnels bored an unprecedented 320 metres in seven days.

Something else that is unique about the TBM is that, as it proceeds on its inexorable way, it leaves a complete tunnel behind. It is not just a very sophisticated drill. Its attendant train is a veritable factory: it disposes of spoil; it receives the lining segments; it puts the segments in place; it injects cement grouting behind the lining; it even carries its own rest room for coffee breaks and its own infirmary, complete with bed, stretcher and equipment for initial emergency treatment. This is another innovation. Any experienced tunneller will tell you that in the past tunnels were built by separate gangs of borers, liners and finishers, working in sequence. It was a muddly process. These TBMs leave behind a 'clean' tunnel.

TUNNEL LINING

Just as geology determines the type of boring machines which can be used, so it also affects the construction and installation of tunnel linings.

The British drives employ curved segments of reinforced concrete. Each is hoisted into position and locked with a final wedge-shaped key segment. The machines are designed for a cycle time of 18 minutes in the service tunnel (7 segments) and 22.5 minutes in the running tunnels (9 segments). The approximately 20-millimetre gap between the lining and the ground is injected with a mixture of pulverized fuel ash (PFA) and cement grouting.

The French undersea tunnels which pass, in part, through more permeable rock, require lining segments of a slightly different construction. Six segments are bolted together to make up a complete ring, and each has its own neoprene gasket to ensure a totally waterproof

seal. The resulting structure is immensely strong and external pressure can only bond the segments more tightly together. Since boring progress is totally dependent on the supply of lining segments, quality and reliability of manufacture and delivery are vital.

The French prefabricated sections are made in a special plant constructed on site at Sangatte. This produces one segment every three minutes and by the end of the project will have supplied a total of almost 400,000 segments. Production at the British end is complicated because of the restricted space at the Shakespeare Cliff site, and manufacturing takes place at a factory site at the Isle of Grain on the north Kent coast. Materials brought there include cement, by rail, and granite and sand, which comes by sea from the west coast of Scotland. Segments are despatched by rail to Shakespeare Cliff round the clock to maintain the supply of over 1,200 per day which the tunnels can gobble up when work is at peak progress.

THE CREWS

The most advanced machine, even when aided by the latest laser and computer technology, is eventually only as good as the men who operate it. For the considerable Tunnel workforce all the best and most experienced crews and technicians had to be attracted for the key jobs. In addition, a huge semi-skilled and unskilled pool of labour had to be found. Nor was it just enough to have the right number of engineers, miners, labourers, electricians, surveyors, caterers, security officers, medical auxiliaries, drivers, secretaries, etc., on site. In conditions where time and efficiency were of the essence and where inexperience or negligence could spell danger, people had to be welded into teams.

The men who hold the top supervisory jobs are members of an international tunnelling élite. They move around the world from one major project to the next and a piece of work such as the fixed link means yet another reunion for Irishmen, Australians, New Zealanders, Sri Lankans, Canadians, Austrians, Welsh, Scots and Englishmen and experts of other nationalities. Peter Bermingham, sitting in the staff bar at the

A bored and lined section of the French landward service tunnel

The tunnellers' shrine near the base of the Sangatte access shaft, dedicated to St Barbara, the patron saint of miners

A roadheader at work. Certain sections of the tunnels were excavated using these machines, including the assembly area below Shakespeare Cliff before installation of the TBMs (left).

Farthingloe camp (known to its occupants as the 'village') where up to 1,000 of the British workforce live, could point round the room to friends he had worked with in Hong Kong, Singapore and Cairo. 'It's a bit of a mafia, an international brotherhood,' he said.

But, as Davie Denman, Agent (that is, Manager) on the marine service tunnel, points out, the workforce is very different to the staff that would normally be found on a tunnelling site: *One thing that makes it so interesting is that it's so big. On a normal job people doing the back-up services and actually working on the machines themselves would be experienced men. Ninety per cent of them are people you take around with you from job to job – experienced tunnellers. Normally you only have to sign a few new guys up. On this job it's the other way round. Probably 10 per cent or 20 per cent of the guys are the experienced men and*

Excavation by hand-held tools of a piston-relief duct in the UK tunnels

the rest of the people have come in from all over the place - bacon slicers from Tesco, lorry drivers, people from other industries. They're all coming in and somehow we have to weld them together.

Half the workforce on the Kent sites are local and, partly because of the high rates of pay the company offers, there was little difficulty in recruiting there. (The taxi driver who took me to the site for one of my visits asked if I could put in a word for him.)

The differentials in Calais are even more pronounced. The key posts are filled by experienced underground workers, some of whom come from other parts of France. However, the large majority of workers (nearly 90 per cent) were recruited locally, and

therefore had to be specially trained. For these workers, the Channel Tunnel is a godsend. Many of them were previously unemployed, some for as long as thirteen years. It is interesting to note that for these reasons wages at the French end of the Tunnel are considerably lower than those being earned in Kent. Nor is there any need for a large construction village in France. The few workers who do not live at home are in lodgings locally.

Somehow these disparate elements gel into teams with identity, a sense of mutual responsibility and a pride in their work. Infectious camaraderie? Shared challenges and risks? High wages? These all help to create a good working environment. Another factor, and

one unique to the project, is its high public profile. The Channel Tunnel has frequently been in the newspapers, unfortunately usually only when something has gone wrong or when a journalist has been looking for a sensational story. Whether it is boardroom rows or allegations of tunnellers' rowdy behaviour that make the headlines, the result is a united sense of outrage:

It was hard because we knew damn different from what the press was saying. We knew how impressive the work actually achieved was. There's a lot of genuine effort and first-class work and it's hard to swallow bad publicity when you're incapable of saying anything in reply.

That was the reaction of Vic Wrightham, Agent

on one of the British land running tunnels, to the newspaper reports that the Channel Tunnel was threatened by squabbles at the top and falling production rates on the ground. His indignation was shared by many.

SPOIL DISPOSAL

As in so many other areas the French and British engineers have had to find different answers to the problems of disposing of spoil. Each had to find a way to get rid of millions of cubic metres of excavated chalk with the minimum of environmental damage. The French had space on land to achieve this. The British did not.

The railway trains that run down the tunnels are made up of flat wagons that convey the lining segments to the TBM and of muck cars that make the return journeys laden with spoil. On the French side, the spoil is taken back to the foot of the main shaft, where the wagons are turned upside-down and the spoil is emptied into a large basin, where huge paddles mix it with water to create a slurry. This is pumped to the surface and up to Fond Pignon, 1 kilometre away. Fond Pignon is an artificial lagoon which has been made by building a 730-metre embankment. By the time the tunnelling is finished, the lagoon will have been filled with 3 million cubic metres of liquefied chalk. Much of the water is pumped away to a purifying plant to be used again. Eventually, the surface will be topsoiled, grassed over and become an indistinguishable part of the landscape.

On the British side, spoil is removed in solid state by railway wagons to the underground chamber at Shakespeare Cliff. It is brought to the surface by a conveyor and deposited behind a new sea wall which has been built to enclose an area of 45 hectares. This has been divided into sections by temporary walls. Each section is filled in turn with spoil. At the end of the operation it, too, will be grassed over.

THE RAILWAY SUPPLY SYSTEM

Each end of the Tunnel will, by the time of completion, be served by about 500 kilometres of temporary railway line. At peak, the British construction operation alone will employ 167

The three-tunnel arrangement
showing cross passages and a piston relief duct
Disposition des trois tunnels montrant des rameaux de communication et un rameau de
pistonnement

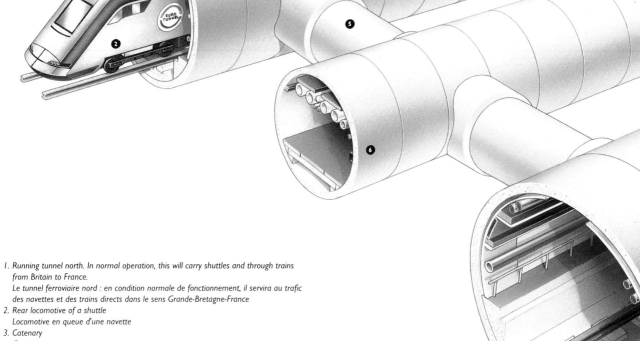

1. Running tunnel north. In normal operation, this will carry shuttles and through trains
   from Britain to France.
   Le tunnel ferroviaire nord : en condition normale de fonctionnement, il servira au trafic
   des navettes et des trains directs dans le sens Grande-Bretagne-France
2. Rear locomotive of a shuttle
   Locomotive en queue d'une navette
3. Catenary
   Caténaire
4. Piston relief duct
   Rameau de pistonnement
5. Cross passage
   Rameau de communication
6. Service tunnel. This will provide an access route for maintenance and, if ever needed,
   access for emergency services and an evacuation refuge.
   La galerie de service : elle servira de route d'accès pour les besoins d'entretien et,
   éventuellement, de chemin d'accès pour les services de secours ainsi que de refuge en
   cas d'évacuation
7. Running tunnel south. In normal operation, this will carry shuttles and through trains
   from France to Britain
   Le tunnel ferroviaire sud : en condition normale de fonctionnement, il servira au trafic
   des navettes et des trains directs dans le sens France-Grande-Bretagne

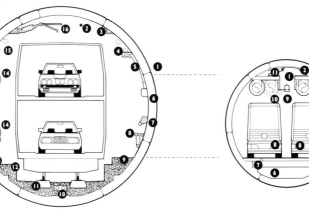

Section through a running tunnel showing proposed installations
Coupe d'un tunnel ferroviaire montrant les installations envisagées

1. Service tunnel side
   Côté galerie de service
2. 25kV feeders
   Câbles d'alimentation de 25kV
3. Traction earth wire
   Câble de terre de la traction
4. Main lighting
   Eclairage principal
5. Cables
   Câbles
6. Guidance lighting
   Eclairage de balisage
7. Fire main
   Canalisation d'eau incendie
8. Handrail
   Rampe

9. Evacuation walkway
   Trottoir pour évacuation
10. Drains
    Système de drainage
11. Concrete rail supports embedded
    in concrete track base
    Supports de rail en béton scellés
    dans l'assiette de voie en béton
12. Maintenance walkway
    Trottoir pour travaux d'entretien
13. Earthing conductor
    Conducteur de mise à la terre
14. Cooling water pipes
    Tuyauteries d'eau de refroidissement
15. Radio antenna
    Antenne de radio
16. Overhead line equipment
    Caténaire

Section through the UK service tunnel showing proposed installations
Coupe de la galerie de service, côté britannique, montrant les
installations envisagées

1. Lighting
   Eclairage
2. Power cables
   Câbles électriques
3. Drainage pipes from drains
   Tuyaux d'évacuation provenant
   du système de drainage
4. Fire main
   Canalisation d'eau incendie
5. Earthing conductor
   Conducteur de mise à la terre
6. Drain
   Système de drainage
7. Buried wire guidance system
   Système de guidage à fil enterré
8. Service tunnel vehicles
   Véhicules de la galerie de service

9. Loudspeaker
   Haut-parleur
10. Radio antenna
    Antenne de radio
11. Power cables
    Câbles électriques

Spoil wagons in the UK north landward running tunnel showing temporary construction tracks

A 'manrider', used for transporting workers along the UK tunnels

A French 'manrider' at the base of the access shaft

locomotives and 866 items of rolling stock. Everything and everyone travelling to and from the tunnel faces must use the trains which trundle constantly along these rails. That means that fresh track must be laid daily, locomotives, trucks and line must be maintained, signals must be kept functioning efficiently. It is a system all to itself.

The construction railway is just one of the services that keeps the TBMs turning. Along the completed tunnel runs an array of cables, pipes and ducts. Power cables feed the electric motors on the TBM, communication cables provide vital radio and telephone links, and overhead there are shielded conductors to energize the massive construction locomotives. A 2-metre duct feeds fresh air to the workforce and on the tunnel walls drainage pipes carry seepage water towards the surface.

THE NERVE CENTRES

The hub of all the tunnelling activity is the underground marshalling area. It is here that

A chef at the Farthingloe village, where steaks are the most popular item on the menu (top left); small-scale tasks are important to complete this large-scale project (top right); tea-break on one of the TBMs (middle left); the early shift (middle right); at the end of the shift a couple of drinks with your workmates (bottom left); French welders (bottom right)

The UK tunnel workers' village at Farthingloe, near the Shakespeare Cliff site (left)

personnel and materials reach the subterranean level and begin their journeys to the rockface, here that the diesel and electric trains disgorge their loads of spoil into bunkers for removal to the surface, and here that the TBMs were assembled before beginning their work.

At Sangatte, the access point is an impressive covered shaft 55 metres in diameter and 75 metres deep. Four lifts and four travelling cranes are constantly in use to carry workers, machinery and lining segments to the tunnel faces. At Shakespeare Cliff more of the marshalling is done above ground at the site at the base of the cliffs. This site is connected not only by two inclined adits to the underground marshalling area but also by an adit which carries road traffic to the Upper Shakespeare Cliff site. A 10-metre-diameter shaft from the administrative centre at Upper Shakespeare Cliff descends directly to the underground marshalling area. If the marshalling area is the hub of each end of the enterprise, the nerve centre is the control room. Highly complex computerized systems feed a range of consoles and displays, which provide a detailed picture of exactly what is happening at every point underground. These ensure the efficient and, more importantly, the safe running of the entire operation. For example, constant measurements fed to each TBM cab and the control room include: the temperature at the cutting head, the humidity level, the oxygen and methane content in the atmosphere, and the efficiency level of the ventilation. Television monitors provide a visual back-up and another display locates every train travelling through the system. Any real or potential danger can, therefore, be located immediately. Men can be warned and evacuated and emergency services directed to the trouble area within minutes.

HEALTH AND SAFETY

Any large work site is potentially a hazardous place. The health and safety of the workforce has been a major concern from the beginning of construction. One factor that complicates the situation is the fact that such a large proportion of the employees were inexperienced and

The Sangatte site, looking across the Channel to the English coast and the White Cliffs of Dover

unfamiliar with a tunnelling environment. To minimize risk all new personnel had to take part in an induction and orientation course before going underground.

Strict safety regulations apply on and around all machines both below and above ground. The Channel Tunnel TBMs are the first of their kind to be fitted with comprehensive fire suppression systems. Heat sensors detect and signal any rapid rise in temperature. The operator can respond either by using hand-held extinguishers or by activating a system which releases foam. As we have seen, the atmosphere in the tunnels is constantly being analysed and information relayed to the control centre. In addition to such measures and all the routine industrial precautions (reinforced by periodic government inspections), all underground workers carry self-

rescuers (carbon-monoxide filters), and a system of tags ensures that every man is logged on and off shift.

Health, of course, is not just a matter of making the working environment as safe as possible. Medical advice has been utilized from the planning stage and right through the construction process. Occupational health provision varies in detail between France and Britain but the standards of care are equally high. Medical examinations are provided, and there is a constant monitor on the well-being of workers and the provision of first aid and safety measures. Medical staff are available to deal with accidents and emergencies, and have also trained many other tunnel workers in first-aid techniques.

In their independent supervisory role the

medical staff are constantly testing and reporting on the working environment. They concern themselves with noise levels, dust analysis, registering of all toxic materials used on site and keeping their own checks on temperature and humidity levels.

Safety is always the prime consideration of men working underground. Quality and progress are important. Competition - between crews and between the workers at the two ends of the Tunnel - is keen. Yet nothing is ever allowed to take precedence over the safety of the operators. Part of the camaraderie of the crews comes from a shared concern for each other's well-being and a determination to reduce potential hazards to a minimum.

Jack-up barges at the Shakespeare Cliff site constructing the sheet piled walls for the spoil lagoons

Spoil disposal at Sangatte. Conveyors take the spoil from the tunnel face along the TBM (top left) and discharge it into wagons, which then return along the tunnel to the base of the access shaft (below). Here the wagons are rotated (next below), emptying their contents into the sump beneath, where the spoil is mixed with recycled water to form slurry. This is pumped up to the adjacent hillside (top middle) to the settling lagoon (top right) at Fond Pignon, a short distance away.

At Shakespeare Cliff, conveyors and wagons also remove the spoil from the tunnel face. Underneath the Lower Shakespeare Cliff site, the spoil is emptied into bunkers (top left) beneath the tunnels. Conveyors then carry it up to the surface and across the site (top centre) where it is deposited in the lagoons by dumper trucks (left and top right).

View of the construction of overbridge number one at the Folkestone Terminal site

CHAPTER 5

TERMINALS

COMINGS AND GOINGS

It has never been done before. That is what you have to remind yourself of when you think about the Channel Tunnel terminals at Folkestone and Coquelles. There is nothing particularly difficult about any one aspect of the construction work but never before have so many aspects been brought together. At each end of the fixed link, road and rail traffic, passenger and commercial vehicles have to be brought together and funnelled into one narrow opening. All the service buildings must be in place to cater for this variety of travellers, who must be delivered to the Tunnel in a strictly controlled, prearranged sequence. At the same time those arriving from the other side of the water will have to be whisked away on to the national road and rail networks with the minimum of fuss and delay.

This involves a concentration of many different kinds of civil-engineering expertise. On different parts of the terminal sites you will find (not necessarily all at the same time) road builders, track layers, bridge builders, tunnellers, drainage constructors, building workers, and the service trades following in the wake of the major contractors. It is not simply a question of all these actors doing their own performance each in their own corner of the stage. There is a producer and a script, and every activity has to follow a defined sequence.

This precise piece of choreography has been conceived and plotted out and should be carried through to execution in record time. Detailed planning began in 1986, and the first shuttles are scheduled to leave the terminals in 1993. Seven years is a remarkably short time for such a large, multifaceted undertaking to run from conception

The terminal site at Coquelles, showing the construction of the overbridges and track formations

The bridges adjacent to the Folkestone terminal site which will carry its road and rail access links

The Coquelles terminal site in 1988, shortly after work began, and (below), another view taken in 1990, showing the progress made in two years

through design and construction to completion. For example, building workshops for the maintenance of railway rolling stock is a straightforward operation - when you know what rolling stock will be coming into the sheds. In 1987 the shuttle designs were scarcely on the drawing-board. They had to pass through several stages before they were approved. Prototypes are still to be tested and final modifications made before the wagons can go into production. Yet, by the time they come into service, the workshops designed and equipped specifically to maintain these vehicles must be up and running.

THE FRENCH TERMINAL

Gerard Vidal, who is Eurotunnel's Site Director in France, analyses the attitude of the French workforce to the Channel Tunnel in these words: 'On each large project the employees are constructors. They build something. But here, in their minds they know that this is an historical project, more than a large project.'

The terminal site at Coquelles dwarfs that and, indeed, all modern civil-engineering projects. Its 700 hectares make it by far the largest land-transport complex in Europe. When finished, its 18 kilometres of perimeter fence will enclose 50 kilometres of railway track, a similar amount of roadway and 26,000 square metres of terminal buildings. Vehicles will enter the site by approach roads from the A26 autoroute and the Rocade du littoral, a new dual carriageway linking Dunkirk, Calais and Boulogne. A system of overbridges and ramps will convey them to eight platforms (planned eventually to be extended to sixteen) from which they will be loaded directly on to the shuttles. There will be a rail freight marshalling yard and, at nearby Fréthun, a high-speed train station.

The major constructional problem was the ground on which all this was to be built. It was marshy and consisted of layers of clay and peat, in some places 12 metres deep. The first task was to compress this into a suitable foundation for building. A 50-centimetre layer of sand was laid to cover the site. Horizontal drainage pipes were embedded in it. Then, a latticework of vertical drains was sunk into the water-bearing strata. Finally, a heavy embankment of spoil was laid (10 million cubic metres of fill in all) to weigh down and compress the original subsoil to create a compact foundation suitable for building and other structural foundations.

From the beginning the French terminal has been conceived of as having a symbolic as well as a functional purpose. That is why 200 of its 700 hectares have been given over to a showpiece leisure and business complex designed to celebrate and display the best of European enterprise. Eurotunnel's brochure calls Le Camp du Drap d'Or 'a city of the 21st century' and promises the very latest in restaurants, exhibitions, shopping and leisure facilities, conference centres and hotels imaginatively laid out with lakes, fountains, parks and boulevards.

THE BRITISH TERMINAL

With only 140 hectares available, the British terminal has had to be planned on a more modest scale. The comparative smallness of the site, its proximity to centres of population and its distance from the tunnel workings at Shakespeare Cliff have created a very different set of construction problems. Environmental issues, in particular, have loomed large at every stage of planning and construction and we shall be considering them in Chapter 6. However, we must mention here some of the engineering challenges which they created.

Drainage is always a major problem on any large site. The original plan was to route the discharge pipeline into a lagoon constructed on the nearby Seabrook Stream. An early ecological survey established that such action could destroy the habitat of a uniquely high number of species of crane-fly. Eurotunnel, therefore, agreed to finance a new tunnel outfall to the sea designed and constructed by the Southern Water Authority.

Walls have had to be constructed to screen the site and building noise from roads and residential areas. Even such a simple process has given rise to interesting technology. Several embankment-retaining walls were constructed using 'reinforced earth' technology, where vertical concrete plates are used in conjunction

The Folkestone terminal site in 1987, looking east (above). Extensive work was required to level the site. The Folkestone terminal site (below) in 1990.

1:20 scale models of the proposed terminals. Overall view of the Coquelles terminal (left) and the central administration block at Folkestone (right).

Levelling the Folkestone terminal site using earthmovers and some two million cubic metres of sand brought in via dredger, barge and pipeline from the Goodwin Sands

with horizontal tie bars to form a stable structure.

In order to cause the minimum disturbance to the landscape, parts of the terminal will be hidden wherever possible. The most impressive engineering work connected with this camouflage is the cut-and-cover process. The arrival loop tunnels, which bring the shuttles in a large sweep to the platforms ready for their return journey to France, and the approach land tunnels between Castle Hill and Sugarloaf Hill, to the west of the terminal, are being constructed using this method. The land is excavated and a solid 'bed' is prepared. On this is placed what is, in effect, a long, reinforced concrete box. Base, walls and top are laid in turn. The sides and bottom can be installed at the rate of 60 metres a week. The top takes longer because it has to be supported on a special structure until the concrete is set. Even so, the rate of progress on these elongated caverns is impressive. Once in place the structures are covered with soil and will eventually be landscaped.

Bringing the tunnels through Castle Hill into the Folkestone terminal requires another technique. This is known as the New Austrian Tunnelling Method. The ground, mainly gault clay, is drilled with a conventional road-header. Then a special concrete mixture, known as 'shotcrete', is sprayed on to the inside of the tunnel using reinforcing mesh and lattice girders to ensure adequate strength. Rock bolts are drilled into the surrounding ground to provide unified arch support. The final process involves lining the inside of the tunnel with 2-millimetre-thick waterproofing lining and then forming the structural lining of in-situ concrete.

Another major problem that had to be solved was how to bring on site vast quantities of fill for levelling, building embankments and creating a suitable building base. Conventional road and rail transport was originally proposed, but then an imaginative alternative, to bring in marine sand from the Goodwin Sands by dredger and pipeline, was adopted. It took ten months to complete all the preliminary negotiations. Various authorities had to be approached and permission for pipe laying had to be obtained

Lining one of the drainage ponds at the Coquelles terminal site

The temporary railhead at Ashford, Kent. Whenever possible, bulk materials are delivered by rail for onward transport by road to the nearby terminal site.

Construction at the French terminal site, showing in the foreground the new bridge built to carry the main coastal road, the Rocade du littoral

Coquelles site with the platform access ramps in the background

from twenty-three landowners. At last, the scheme was agreed and the complex operation could begin. Dredgers brought the sand to a floating pumping station moored 700 metres off Seabrook beach between Hythe and Sandgate. There, it was mixed with sea water in a 1:4 ratio and pumped through steel pipes across 5.3 kilometres of open country, roads and a railway line. The mixture arrived in two large reception lagoons from which the water drained into a third and was pumped back to sea along the same pipe at regular intervals. This process delivered 2 million cubic metres of fill to the site in nine months, and had the major advantage of saving an estimated 200,000 local lorry movements.

As with the Coquelles terminal, the main problem lay not in solving any of the individual engineering problems, innovative though some of the solutions were, but in integrating every aspect of the work into an overall programme. It is not technology which, in the last analysis, is the biggest Channel Tunnel miracle. It is logistics.

Testing the first tracks laid at Dollands Moor, where British Rail's sidings will be located. It is near this point that the continental main line will diverge from the existing London-Folkestone route visible on the left of the picture

This wall was built along the southern side of the Folkestone terminal site to retain the sand infill that will provide a level base for the terminal (right)

View of the Coquelles site, with part of Calais in the background to the right

Aerial view of Holywell Coombe with the Folkestone terminal site behind

Construction of the cut-and-cover tunnels through Holywell Coombe

The TBMs boring the landward drives from Shakespeare Cliff emerge at Holywell. Connecting these tunnels to the terminal site involved the construction of cut-and-cover tunnels through Holywell Coombe and excavated tunnels using the New Austrian Tunnelling Method under Castle Hill. The cut-and-cover technique requires the cutting of a trench in which a boxed section tunnel is constructed. This trench is later covered and landscaped. A roadheader machine is used for excavation by the New Austrian Tunnelling Method. Steel lattice girders, sprayed concrete and steel mesh provide initial support. The tunnel is then lined with butyl before a permanent concrete lining is constructed in situ.

At Castle Hill, tunnelling methods changed from cut and cover to the New Austrian Tunnelling Method, the beginnings of which can be seen in the background

Replacing the teeth on one of the roadheaders used to excavate the tunnels through Castle Hill (right)

PVC liner being fitted

Support meshing over the liner before the permanent concrete lining is installed

Chalk downland area near Shakespeare Cliff, looking across to the French coast

CHAPTER 6

ENVIRONMENT

THE EGGS AND THE OMELETTE

In the late 1980s we all went 'green'. It came home to most of us in the First World that we are endangering our own habitat and that the price we are paying for our increasingly sophisticated materialistic culture may be too high. Now we are faced with a dilemma: technological advance is essential to the survival of our highly complex, industrialized societies but, if the quality of life is to be safeguarded, we have to protect our world. Nor is it simply a matter of choosing between 'progress' and 'environment'. The two go hand in hand. In south-east England, for example, whether there is a Channel Tunnel or not, transport requirements will increase. That means either congestion or improvements and extensions to the road and rail network. Either way, the environment and the quality of life are under threat. That is why in major development projects such as the building of the Channel Tunnel there has to be the closest possible consultation between the developers, constructors and environmentalists, from the planning stage through to completion - and beyond. In fact, as part of the documentation required in the competition for the various fixed-link proposals, the two governments required an environmental impact assessment to be done. This was the first assessment to be prepared under the new draft directive issued by the European Commission.

STATUTORY OBLIGATIONS
The French and British governments place considerable constraints on developers. Every area has its local plan, indicating where and under what conditions future construction may be

permitted. Beyond that, government legislation protects the interests of residents and other interested parties. For example, in the UK, under the terms of the Ancient Monuments and Archaeological Areas Act, 1979, no planning consent is granted within an area of archaeological importance until an archaeological survey has been made and plans agreed either not to disturb important sites or to preserve them by record (which usually involves excavation). Similar legislation exists for the preservation of flora and fauna.

The French also possess conservation laws that make provision for more central control.

A rare Late Spider Orchid (Ophyrs holoserica) growing on chalk grasslands

Although considerable powers rest with local authorities, the concept of *aménagement de territoire* (overall land management) means that when a project has national significance a public inquiry tries to reconcile the wider benefits with local environmental issues. Naturally, the vastly different population patterns in Kent and Nord-Pas-de-Calais mean that environmental problems have always been more acute at the English end of the Tunnel. Under the terms of the Channel Tunnel Act, the local planning authorities, although they could not prevent the building of the fixed link and its terminals and related transport systems, have had considerable powers in shaping the development in detail. For example, the Act requires Eurotunnel to agree specific statements on minimizing disruption caused during the construction period by noise, dust, the transportation and storage of bulk fill materials by road and rail, as well as night-time working. A landscape scheme for each of the permanent works as well as for the

Control room of the dredger Barent Zanen

Above: Water spraying at the Folkestone terminal site to damp down construction site dust

Left: A settlement lagoon receiving a sand/water mixture. By pumping this in through a pipeline many thousands of lorry movements were avoided.

Monitoring the marine environment

'Well, it wasn't there when I put the cat out last night!'

Cartoon from Chartered Surveyor Weekly, 23 June 1988

The villages of Newington and Peene, next to the Folkestone terminal site. Measures were taken to minimize disturbance: for example, a bund was installed (top right) to reduce noise and visual disturbance.

reinstatement of temporary work sites also had to be agreed.

Before granting approval, the local planning authorities were required by the Channel Tunnel Act to consult fully with the residents of the area and also with various statutory bodies including the Nature Conservancy Council, the Countryside Commission, the Historic Buildings and Monuments Commission, and the National Rivers Authority (previously the Southern Water Authority) where any of the requests for approval included matters in which these bodies are interested.

ENVIRONMENTAL STUDIES

Colin Kirkland, Eurotunnel's Technical Director, who is ultimately responsible for all environmental issues, believes that the company has gone far beyond its strict statutory obligations: 'One of the reasons I believe we won the concession in the first place was because we did a more detailed environmental impact analysis than any of the other bidders'. Eurotunnel's scheme was the first to set itself the

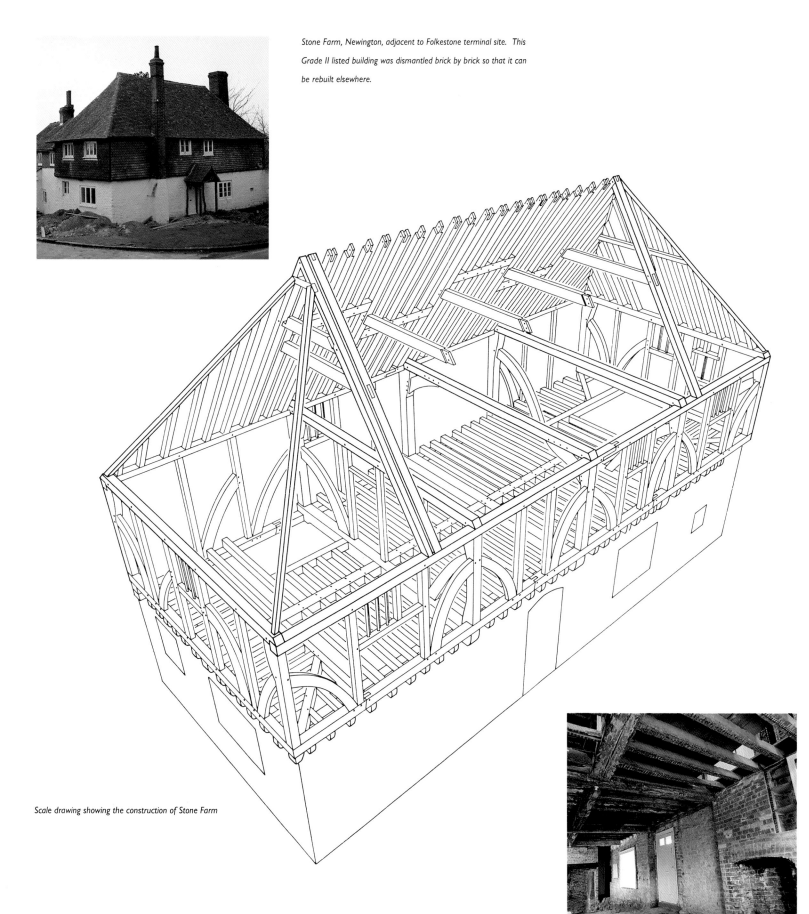

Stone Farm, Newington, adjacent to Folkestone terminal site. This Grade II listed building was dismantled brick by brick so that it can be rebuilt elsewhere.

Scale drawing showing the construction of Stone Farm

Interior view of Stone Farm taken during the dismantling process

Excavating a Roman villa near the Coquelles terminal site

Geological and archaeological investigations carried out at Holywell Coombe before construction work started

standards of the European Community's Directive on Environmental Impact Assessment, then in draft form. This document sets out strict requirements for the detailed investigation of the impact of both the construction and operation of major projects, such as the Channel Tunnel, on the local environment. Eurotunnel commissioned eighteen specialist reports on the land and marine environment and the likely effects of the tunnel and terminal construction operation. They dealt with bats, grassland renewal and a host of other topics. These initial reports provided information on the existing environment, the likely effects of the development and possible methods of mitigating any adverse effects, and have been followed by detailed baseline studies so that any changes during construction and afterwards can be monitored.

All this was not done out of sheer philanthropy. Eurotunnel is, of course, a commercial undertaking, formed to build and operate a transportation system and show a profit. But nor can these environmental concerns be put down to cynical manipulation of public opinion. Colin Kirkland admitted that, during the parliamentary committee stage of the Channel Tunnel Bill,

a lot of environmental questions were raised by individuals and authorities and we didn't have time to argue them at great length, because we were up against a time schedule for completion. So we gave commitments to do things which, I suppose, if we'd had unlimited time we might have argued more about. But I'm very pleased about that because it means we've managed to create a good reputation for a project which must unquestionably disrupt large tracts of land. And at the end of the day I think we shall have done things which will be better for the environment that any of the other cross-Channel fixed-link proposals, which would have had a much more serious effect on Kent and Nord-Pas-de-Calais.

Colin is convinced that, managed properly, the Tunnel and its transport links will be better for the environment than any system which would be forced on Kent and Nord-Pas-de-Calais by alternative development.

CONSULTATION

The consultation process in France did not take long. The actual area of the construction sites was very sparsely populated. A handful of farmers received compensation from the government. Most inhabitants of the wider area had little reason to protest against the building of the Tunnel and every reason to be in favour of it. It would bring to their region much-needed employment, new commercial, industrial and leisure facilities, and a greatly improved transport system.

The British end of the Tunnel debouches into a well-populated coastal strip bordering the Downs and the Weald of Kent, the 'Garden of

Eurotunnel has supported local groups seeking to protect local wildlife: catching toads on a road near their spawning ground (left); newt catching before construction began (right)

Transplanting Biggins Wood from the Folkestone terminal site. This 'transplant operation' is an attempt to preserve the flora of this small remnant of ancient woodland.

The escarpment above the Folkestone terminal site. This area of chalk grassland forms part of the White Cliffs Countryside Project and in the future will be preserved for grazing.

Anti-Tunnel protesters in Folkestone, opposite Eurotunnel's first information centre at Tontine House

England'. The consultation process there was certain to be more protracted and more sensitive. It had two aspects: the official deliberations with 120 local authorities and significant numbers of interested groups, and the informal contacts between Eurotunnel and local people.

The official consultations channelled recommendations to Parliament during the Committee Stage of the Bill and subsequently to the planning authorities. As a result, several amendments were made to the original design. The biggest was changing the access to the terminal. As originally conceived, this would have come in from the north-west, disturbed two local communities, and taken more land out of agricultural use. The revised scheme brings the access in from the south-west and is generally thought to be less disruptive.

Most people involved in official bodies seem to have been reasonably satisfied with the consultation process. Bob Summers, Chief Executive of Shepway District Council, reckons that he and his colleagues petitioned on thirty-three important points and received satisfaction on thirty-two. Peter Raine of the Kent Trust for Nature Conservation commended the developers for genuinely caring about the environment and being prepared to go beyond their statutory responsibility. He cites the Etchinghill Escarpment, a Site of Special Scientific Interest (SSSI). This rich grassland system provides a habitat for several rare butterflies, orchids and other plants. The area has not been disturbed by site works but, nevertheless, Eurotunnel have fenced and grazed the area in order to bring the land back into good heart.

All the official bodies approached their work understanding clearly that the Tunnel was going to be built and that, therefore, their task was to minimize any damage and maximize any benefits that might come from it.

Several members of the general public took a more radical view, at least in the early stages. Some faced the prospect of having their houses and land compulsorily purchased, others of seeing the value of their property diminished, of having to put up with seven years of disruption and inconvenience and seeing their traditional way of life permanently changed. It is not surprising that there was a vigorous anti-Tunnel

Eurotunnel's Exhibition Centre near the Folkestone terminal site, visited by over 600,000 people in its first two years

lobby. In February 1986, only three weeks after the French and British governments had announced that the Tunnel was to go ahead, Eurotunnel opened a small information centre in Folkestone. Detailed information was provided about many aspects of the project, and local residents were encouraged to come and see for themselves and discuss the likely impact of the Tunnel project on their lives or property. Regular meetings were held with representatives of local groups, and the centre undoubtedly helped people to understand the project better; it also helped them to identify their particular concerns more accurately during the select committee hearings.

The completion of the parliamentary proceedings, and then the completion of Eurotunnel's fund-raising exercise in autumn 1987, enabled construction to start. The project's impact on local residents was now much more immediate and direct.

There were two ways of dealing with genuine and imagined grievances. The first was the appointment of an independent ombudsman. This was the former British Ambassador to Sweden, Sir Donald Murray, KCVO, CMG. He was appointed Channel Tunnel Complaints Commissioner and his office in the centre of Folkestone was available to receive calls from the public. As a distinguished public servant and a

long-term resident of Kent, Sir Donald was much respected and was able both to explain when things could not be changed and to ensure that when necessary the changes *were* made. In 1988 Sir Donald dealt with 380 issues raised by members of the general public, ranging from damage by lorries to interference with television reception. In 1989 the number of complaints had dropped to 260 and it continued to fall during 1990.

By the time people reached Sir Donald they were often angry or, at least, very worried. Eurotunnel realized that it needed to reach the public before that annoyance level was reached and explain what it was doing. This task fell

Public consultation meeting held by Eurotunnel. From the outset it was Eurotunnel's policy to keep local people informed about the project.

largely to Eurotunnel's staff in Folkestone, and in particular to Penny Smith and to Kent Public Affairs Manager Tony Gueterbock, who has been with the project since its very beginning. Later, Eurotunnel's environmental manager Elisabeth Culbard, and Katharine Kershaw, her opposite number with the contractor TML, joined the team. Someone was always on hand to deal with complaints. The early days, with an average of twenty-eight serious complaints a month, were, as Dr Culbard recalls, 'extremely unpleasant for those people who had to answer questions'. Penny Smith remembers that 'we had all sorts of vandalism: bricks through the windows, paintings on the wall of the car park - even a sheep's eye

stuck on a video!' Local residents were understandably angry, initially at the prospect of the future project, later at the disruption it caused: mud on the roads in wet weather, and dust in their houses in dry weather, and noise in all weathers. The staff worked hard - listening patiently, explaining, relaying genuine grievances, improving situations where they could. For example, as we have seen, when the level of the terminal site had to be raised, 2 million cubic metres of sand (transported from Goodwin Sands) was brought by pipe from Seabrook, thus saving an estimated 200,000 truck movements. To reduce the impact of noise, Eurotunnel provided noise insulation to all houses where the

construction of the terminal was estimated to exceed the specified noise limits. Residents who felt unable to carry on living in proximity to the terminal were able to sell their houses to Eurotunnel at a fair market price. The policy paid off. As the local population came to understand what the developers were doing and saw that they were dealing as effectively as they could with genuine grievances, the Channel Tunnel became something local people learned to live with.

The information centre at Folkestone helped this process. From the start, it attracted a lot of interest from visitors and school parties, in addition to local residents. Soon it became clear that the construction of the Tunnel would attract great numbers of visitors, and local authorities and residents were concerned that they would clog the very narrow local roads. Thus, during a select committee hearing, Eurotunnel made a commitment to construct a viewing platform for the public. Subsequently they decided to incorporate the platform into a much larger and exciting exhibition centre situated just across the M20 from the Folkestone terminal site.

The move to the new building took place in September 1988, and the Exhibition Centre is now a major tourist attraction, welcoming over 600,000 visitors in its first two years. It also maintains its original function as a place for information and discussion of local issues, complaints and so forth.

ARCHAEOLOGY

The country on each side of the Channel has been continuously occupied for over 12,000 years and is, therefore, immensely rich in archaeological content.

After the initial survey in Kent, excavation and other kinds of archaeological investigation continued on the terminal site and the adjacent area before and during the construction activity. The work was carried out for Eurotunnel (and at their expense) by a team from the Canterbury Archaeological Trust under the direction of Jonathon Rady. They certainly had a tough time, working through one of the wettest winters on record and the 1987 hurricane which flattened all their huts and removed one without trace. The finds were not dramatic but the exercise was valuable, and probably unique in enabling a picture to be built up of the occupation of such a large area over several centuries. Very little work had been undertaken there before, so the Eurotunnel commission provided a welcome opportunity to add to knowledge of Kent's past. The results will keep archaeologists busy for several years, cataloguing, describing and assessing the significance of the 30,000 pottery shards, three burials, jewellery, an Iron Age sword and other artefacts gathered from 13 kilometres of trenches.

The earliest societies of which evidence emerged were the 'poor relations' of the day. During the Bronze and Iron Ages the wealthier and more powerful Britons occupied the salubrious, hilltop sites. The lowlands were occupied by the 'overspill' - pastoralists and subsistence farmers whose passing leaves little trace in the ground.

In France the area at the disposal of archaeologists was very different. A small team from Lille was brought in, funded by Eurotunnel, the government and the local authorities. They had the much larger terminal site to explore and had to adopt a selective approach. What they revealed was not the relics of a depressed, slave society but occupation sites where, over the centuries, the more affluent inhabitants of Gaul had lived. They made some magnificent, rich discoveries including Bronze Age ring ditches, a Roman villa, and a medieval church and cemetery.

NATURE CONSERVANCY

Eurotunnel is responsible, not only for landscaping the terminal sites and the other construction sites, but also for ensuring that the surrounding countryside suffers as little as possible. The biggest change resulting from the work will be the platform at the foot of Shakespeare Cliff, built up with excavated spoil. This 'false shore' interposed between the sea and the 'white cliffs' will house buildings associated with the permanent cooling and ventilation systems for the Tunnel. It will be landscaped and part is planned to be designated as a nature reserve or public amenity area in accordance with plans drawn up with the local authority. One of the 'eggs' that had to be 'broken' to make the Channel Tunnel 'omelette' was Biggins Wood. This 5-hectare remnant of ancient woodland was in the centre of the Folkestone terminal site, so Eurotunnel moved it. Seeds and soil were taken to a specially constructed nursery, from where the resulting plants will be relocated on new sites near the terminal. It is interesting to note that earlier developments for road and house building would simply have destroyed part of Biggins Wood.

These are only two of several projects in which the developers are engaged. More all-embracing is the White Cliffs Countryside Project. Eurotunnel is participating with local authorities and conservation groups in a project to protect and enhance the chalk grassland not only around the construction sites but over a much wider area of east Kent, including some 28 kilometres of coastline and running some 16 kilometres inland. The work will involve reinstating the chalk grassland, removing eyesores, providing footpaths, signposts, car parks and other public amenities.

PEOPLE

In Kent the emphasis has been on disturbing as little as possible the way of life of the people already living in the area. In Nord-Pas-de-Calais the inhabitants are looking for enhancement of life rather than the preservation of their existing standards.

In 1520, two extrovert monarchs, Henry VIII of England and Francis I of France, met near Calais for one of the most extravagant contests of diplomatic one-upmanship ever staged. At the Field of Cloth of Gold, Le Camp du Drap d'Or, each tried to outdo the other in feats of arms, personal adornment and costly magnificence. The modern Camp du Drap d'Or, which will form the Coquelles terminal site, will lack the element of rivalry but will live up to its name in terms of the splendour of its conception.

Le Camp du Drap d'Or will include a 200-hectare business and leisure centre. By designating the site a *Zone d'Aménagement Concerté* (ZAC), the French government expressed its intention to ensure that the terminal was a showpiece for the region, a magnet for investment and employment and a means of encouraging industrial and commercial initiative. What is of more immediate concern to the people of Nord-Pas-de-Calais is that something in the region of 13,000 to 15,000 jobs, of all kinds, are expected to be created by the terminal and the Camp du Drap d'Or.

The massive model layout (1:20 scale) of the entire Tunnel project on display in Eurotunnel's exhibition centre at Sangatte. A similar model layout is on view at the Folkestone Exhibition Centre.

Artist's impression of the proposed business, leisure and cultural park to be built adjacent to the French terminal site

View of the Field of Cloth of Gold where Henry VIII of England and Francis I of France met in 1520. Now Eurotunnel is proposing a 200-hectare cultural, leisure and communications centre with the same name.

The TGV atlantique, which links Paris to Brest and to Le Croisic at speeds of up to 300 km/h

CHAPTER 7

THE FUTURE

A TRANSPORT SYSTEM FOR THE TWENTY-FIRST CENTURY

For France what the tunnel means is the opportunity...of making TGV nord the paying proposition which all parties in Nord-Pas-de-Calais want...TGV nord involves a new track, a new line, the servicing of several towns and the development of new techniques...Nord-Pas-de-Calais has suffered a great deal in recent years from the decline of the coal, textile and construction industries. By contrast, Kent is a superb area - prosperous and with no unemployment problem. Consequently, environmental issues are very important in British eyes. So, we have two completely different regions.

That was how André Bénard, Chairman of Eurotunnel, summarized the contrasting problems and perceptions of the two areas which will be most affected by the Channel Tunnel, in an interview for *Sphères*, the journal of the Crédit Lyonnais in 1987. Given the major economic and social contrasts between Kent and Nord-Pas-de-Calais, it is understandable that there should have been different feelings on each side of the water about the prospect of a growing volume of through traffic. But what changes will the Tunnel bring *in reality* to the two areas most concerned?

THE FRENCH HINTERLAND

If past experience is anything to go by, the completion of the high-speed rail link from Paris to Calais should have an enormous impact on the French coastal region. SNCF, the French railway company, has invested heavily in the *Train à Grande Vitesse* (TGV), the jewel in its crown.

The shopping and business heart of Lille, one of the cities that stands to gain from commercial developments stimulated by the Tunnel

T.G.V. NORD, TU PASSERAS PAR AMIENS OU TU NE PASSERAS PAS!

Postcard issued by Amiens, which campaigned to have the TGV nord route pass through the town

And to good effect. The high-speed passenger train has enjoyed a success far exceeding the hopes of its designers. In 1981, the TGV sud-est line from Paris to Lyon was inaugurated. Trains hurtled along the specially constructed line at speeds of up to 260 km/h, almost halving the previous journey time of four hours. In the first three full years of operation rail traffic increased by 147 per cent. Much of the increase was accounted for by travellers switching from air to rail, but even so there was a clear growth in the total number of journeys made. Improved communication led directly to the new commercial developments in the area served by the new line. In September 1989 the first part of the TGV atlantique line was opened, from Paris to the west coast, and cities such as Rennes are already planning new commercial developments. Christian Baeckeroot, a member of the French

The open road in France: relatively empty in contrast with the congestion on the British side of the Channel

Parliament, hit the nail on the head when he said, during a visit to Eurotunnel's Calais site in 1987, 'In my opinion what matters is not so much the Tunnel, as what will happen to the infrastructure on both sides of the water'. Regional politicians and businessmen are looking to the new road and rail links to bring new commerce and industry to a region which lost 135,000 jobs between 1975 and 1984, and plans are underway to maximize the new opportunities. At Calais a 700-hectare commercial and industrial development zone is being built. In Lille the local authority is providing grants for the clearance and reclamation of land so that new factories and office complexes can be constructed. And the city has contributed 900 million francs towards the cost of a TGV interchange station.

The future may look very promising for towns close to the planned TGV lines and the new

autoroutes, but what of those further away from the new transportation arteries? Roads and rails that rush people and goods to one town can just as easily suck enterprise away from conurbations too distant to benefit directly. The people of Amiens were up in arms when they realized that the TGV nord line was to be routed through Lille and would by-pass the capital of Picardy by 40 kilometres. They formed the Amiens-Picardy TGV Association to campaign for the rerouting of the line. The Association used a variety of methods ranging from conventional lobbying in Paris to buying a 15-hectare plot of land on the rail route, splitting it into thousands of portions and selling them to supporters - thus forcing the railway company to enter into protracted, time-consuming negotiations. So far they have only extracted from the government a promise to consider a new high-speed Paris-Amiens-Calais

line at some future date. The Amienois will, however, be at the centre of an improved road network; autoroute A16 will link at Amiens with upgraded highways N1 and N29 to Calais, Le Havre and Rouen.

Another potential problem was the fate of the existing ferry ports, Boulogne, Calais and Dunkirk. The search for new business could have resulted in a desperate cut-throat rivalry between the three harbours. But thanks to the initiative of Michel Delebarre, France's Minister of Transport and Mayor of Dunkirk, this has been averted. The three ports are to be drawn gradually into a corporate entity which will channel investment, plan joint commercial policy and harmonize tariffs. The object is not simply to survive the loss of traditional trade but to exploit fully the new prosperity coming into the area.

Satellite view of Britain, Ireland and part of Europe. The Channel Tunnel seems set to play a key role in stimulating commercial and leisure traffic within the European Community.

traditionally popular with the British. And along the coast, a government initiative called Mission Côte d'Opale is helping to attract investment in hotels and leisure and sporting facilities. Nor are the possibilities of cross-Channel holidays in both England and France being forgotten. Nord-Pas-de-Calais and east Kent are already being promoted as a single holiday area, and their potential for attracting customers will increase when the Tunnel opens.

THE ENGLISH HINTERLAND

Kent, as we have seen, does not want and could not cope with massive industrial investment. The scope for factories (even high-tech ones), transport depots and warehouses is very limited. The county is not a traditional manufacturing region looking to be revitalized. In the west and some central parts it is largely commuter and retirement country, and people are understandably very anxious to preserve the peaceful countryside and pleasant country towns. In east Kent (the area around Folkestone, Dover and up as far as Margate), there is some small manufacturing, and in addition many people work in tourism and other service-sector jobs. Unemployment in east Kent has traditionally been high; in July 1990 it was 2,091 in the Folkestone area, rising to 3,582 in Thanet. The number of people unemployed has fallen considerably in the last few years, partly as a result of the Channel Tunnel construction. However, by 1993 the 7,500 jobs created by the Tunnel construction will come to an end, and Eurotunnel's future transport operation will create many fewer new jobs. Similarly, the cross-Channel ferries are also rationalizing their operations and, in addition, the Single Market is likely to cause a significant reduction in the freight-forwarding business.

On the positive side, the Kent Impact Study (an economic study undertaken by government, local authorities and Eurotunnel) defined a number of areas where significant new job opportunities could be created, both as a result of the opening of the Channel Tunnel and of the Single Market. The challenge to the Kent hinterland is to find ways of extracting the

While Kent forms one of the main British transport corridors to the Continent, there is much beautiful countryside to be preserved

The French game of boules being played in an English pub garden

Tourism in Nord-Pas-de-Calais should also benefit from the Tunnel, but only with careful planning and considerable investment. At the moment, it is not a significant holiday area, either for the French, who tend to go south on holiday, or for the British. The Tunnel will make the whole of the Continent more accessible to British travellers. The aim of the Regional Tourism Committee is to persuade some of these to stop off and explore Nord-Pas-de-Calais, which, in the jargon of the trade, it is developing as an attractive 'budget' destination, especially for weekends and short-break holidays. Self-catering accommodation is being developed. The extensive inland waterway system is being reopened to provide boating holidays,

Dover Harbour, the ever-busy departure and arrival point for travellers crossing the Channel

The new generation of cross-Channel ferries includes the Pride of Dover

The Sea Cat, a mammoth catamaran being introduced on the Portsmouth-Cherbourg route

maximum benefit from the Channel Tunnel.

One such is undoubtedly tourism. Until recently, Dover had little to offer to tourists, beyond providing basic facilities (hotels, petrol, meals, etc.) for travellers on the cross-Channel ferries. Now all that is changing rapidly, and the enormous historic interest of the town, which has played a continuous and crucial role in the nation's story since Roman times, is belatedly being recognized. The White Cliffs Experience, a heritage exhibition using the latest audio-visual techniques, is the focal point of the new developments, which are aimed at attracting visitors to the entire east Kent coast.

Folkestone, by contrast, is concentrating rather more on residential and leisure developments. Linda Cuffley, Chair of the local District Council, believes that the high-speed rail link, still under discussion, will be crucial. If it is built, she believes it will make Folkestone 'an easier and more attractive place to live and commute, and a place to station yourself if you want a foothold in this country but you also want to be in touch with the Continent'.

The town that is likely to derive most direct benefit from the Tunnel is Ashford. Not only does it lie a few miles north of the Folkestone terminal on the direct road route to London, it also hopes to have an international passenger station, at which some cross-Channel trains are planned to stop, and a road freight clearance depot. All this should make it a natural centre for commercial and industrial concerns requiring a base within easy reach of the Continent. In the wake of new business will come services such as conference centres, hotels and shops.

THE NEW EUROPEAN MARKETS

In 1973 Great Britain joined the European Economic Community. Since then its trade with the Continent has more than doubled. In the same period, cross-Channel traffic has more than doubled. Continental leisure travel (including day and weekend trips) is becoming more and more frequent. Gradually, the British are turning away from their traditional trading and cultural links across the world, especially with the Commonwealth and the United States, and are

Flag of the twelve-nation-strong European Community

facing toward the European mainland. The Single European Market, which will come into effect on 1 January 1993, is helping the process. Goods will flow faster and more freely across borders, companies will find it easier to establish operations in other countries.

Europe itself is changing as well. All twelve European Community (EC) countries will be involved in the Single Market, and other countries (e.g., Austria and Turkey) are bidding for a closer association with the EC. Vast new markets are opening up in Eastern Europe. With a population (324.5 million in Western Europe, another 137.5 million in the previously Communist countries of Eastern Europe) far greater than that of its main competitors (250 million in the USA and 130 million in Japan), the new Europe could become one of the world's most powerful economic forces.

The economic heart of Europe - the so-called 'golden triangle' - embraces northern Germany, the Low Countries, south-east England and northern France. The Channel Tunnel will have a vital role in developing communication and business at the heart of this golden triangle.

The potential benefits for UK-based commercial enterprises are greater than those for their mainland competitors. British businessmen will have improved access to the vast European market, since all UK exports to the Continent have to cross the Channel. Their counterparts in Brussels, Lyon and Frankfurt, by contrast, will merely find it easier to reach the 50 or so million customers in the UK: far from insignificant, but scarcely comparable. Varying forecasts have been made about the increase in cross-Channel traffic during the 1990s and on into the twenty-first century. Much of this

increase is normal growth (that is, it would happen anyway, even if the Tunnel were not being built). Some of it is so-called 'created' (additional) traffic, and part of this, the experts consider, will be the result of the opening of the Tunnel. All categories of traffic - road and rail, passenger and freight - seem likely to increase, and forecasters suggest that levels in the year 2013 will be approaching twice those in 1993.

THE TRANSPORT INFRASTRUCTURE

All this means that a vastly improved transport infrastructure is necessary. When the Tunnel opens, much of this infrastructure should be in place on the Continental side of the Channel. TGV nord, linking Calais, Lille and Paris, should recently have been completed, and TGV est, from Paris to Strasbourg, is also timetabled to be nearing completion. These lines are planned to link with those of other countries. For example, a committee of fourteen national railway companies is working on the creation of a high-speed Europe-wide rail network planned to be in place by the end of the century. Three major roads will connect Calais with Paris, Dunkirk and Ostend, and the coast south beyond Boulogne, and these will feed traffic into the already existing motorway network of France and Belgium. The Tunnel has been a considerable stimulus to all this activity, but much of it would have happened anyway.

The big question is whether Britain will be connected effectively to this system. English roads and railways suffer badly from congestion, especially in the south-east. Any road traveller who crosses the Channel is immediately aware of the contrast between the crowded roads threading the suburban sprawl of the Kent and Sussex coasts and the straight, empty roads of northern France. The Tunnel, therefore, presents British planners with a major challenge - to ensure that road and rail traffic flows as swiftly on their side of the Tunnel as it will on the other. If they fail to meet that challenge, the benefits of the fixed link will not be fully enjoyed by the British, least of all in the regions.

The problem breaks down into two parts: getting traffic from Folkestone to London, and

The M25 London orbital motorway

A crowded train arrives at Charing Cross station in London's West End

getting through traffic beyond the capital and on to the national network. The M20 motorway is planned to be complete by June 1993, linking the terminal with the M25 London orbital road, which will have been widened to four lanes in some places. Conventional railways will have been improved between Ashford, with its international station, and Waterloo, the London rail passenger terminal. British Rail has acknowledged that ordinary commuter traffic on the already congested Southern Region is continuing to increase. However, there seems to be little prospect of a high-speed line being built much before the turn of the century. Although Kent County Council has expressed support for the concept of the high-speed line, considerable problems of promoting, constructing and operating it through Kent and the London suburbs without any government financial aid remain.

On the nationwide scale, BR has announced its intention of running, in conjunction with the French and Belgian national railways, a typical summer service of fifteen international passenger trains a day in each direction between London and Paris, and the same number each day in each direction between London and Brussels. Additional services, running on existing tracks, are planned to run from Manchester, Wolverhampton, Edinburgh, Glasgow, Leeds, Swansea and Plymouth. In the long term, a new

international passenger terminal at King's Cross in central London, linking the fast main lines to the Midlands and the north of England with a new high-speed rail link to Folkestone, could allow a very effective service to be provided from the regions as well as from London. The new passenger rail link would carry both international trains and services for commuters from east Kent, thereby releasing sufficient capacity for freight trains on the existing lines. On the freight side, daily services are planned from a network of regional terminals to the major industrial centres of the Continent. BR expect that over 70 per cent of this international freight traffic will begin or end its journey beyond London, and claim that 400,000 lorry trunk hauls a year will transfer to rail.

Many people in business, as well as transport lobbyists and other interest groups, have been clamouring for years for an integrated British transport policy. The opening of the Channel Tunnel makes such a policy even more urgent. In 1985, SNCF reached a major agreement with the French government whereby it could spread out existing debt repayment and plan for the future with state aid. No such agreement exists in Britain. Indeed, the terms of the Channel Tunnel Act specifically preclude the government from giving financial support to British Rail's Channel Tunnel services. The government's refusal to move from this position in relation to the new rail link puts Britain at a disadvantage in relation to its Continental competitors. As Alastair Morton, Eurotunnel's Chief Executive, among many others says, 'If the government will not commit the resources to permit even competition with Continental members of the Single European Market, how likely are they to accept the more difficult adjustments?'

VEHICLES PER KM

6,853	3,974	3,458	2,141 (1986)
UK	FRANCE	WEST GERMANY	USA

Traffic density, vehicles per kilometre of motorway, from 1988 data supplied by the British Road Federation

A model of the 'Three Capitals' train, which is planned to travel to London, Paris or Brussels and significantly reduce existing journey times

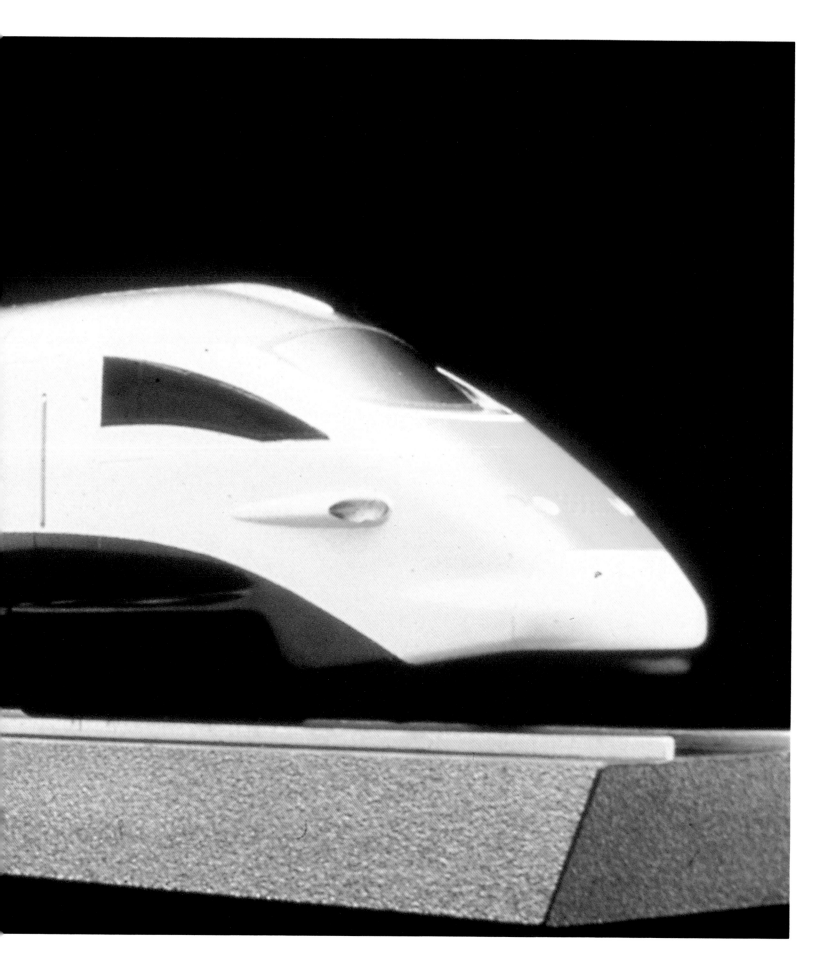

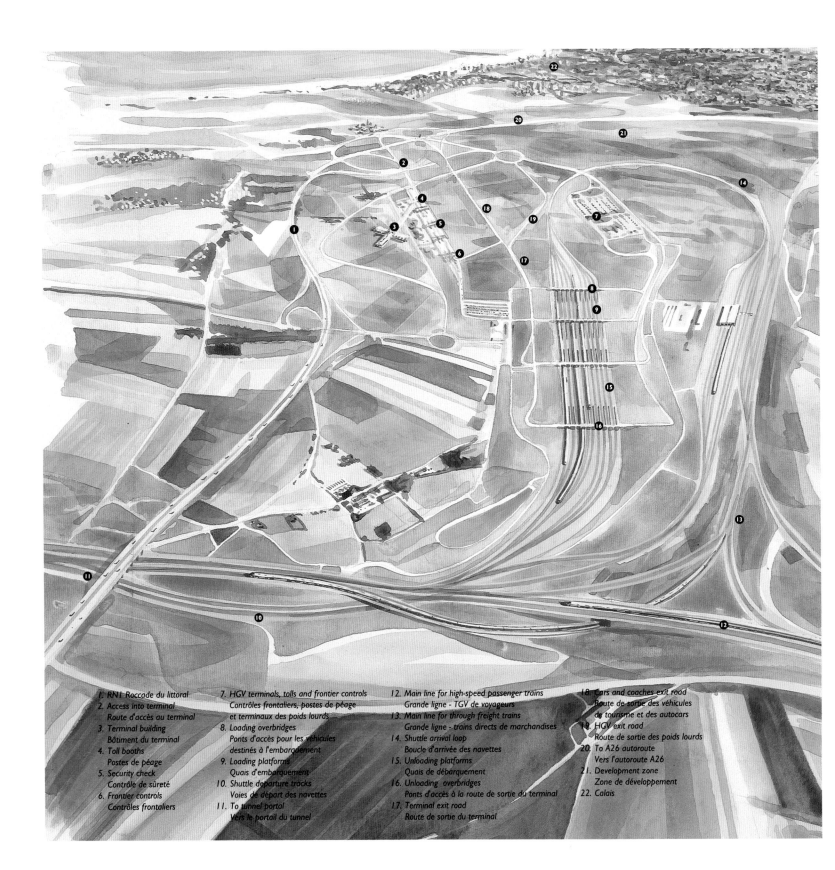

1. RN1 Rocade du littoral
2. Access into terminal
   Route d'accès au terminal
3. Terminal building
   Bâtiment du terminal
4. Toll booths
   Postes de péage
5. Security check
   Contrôle de sûreté
6. Frontier controls
   Contrôles frontaliers

7. HGV terminals, tolls and frontier controls
   Contrôles frontaliers, postes de péage
   et terminaux des poids lourds
8. Loading overbridges
   Ponts d'accès pour les véhicules
   destinés à l'embarquement
9. Loading platforms
   Quais d'embarquement
10. Shuttle departure tracks
    Voies de départ des navettes
11. To tunnel portal
    Vers le portail du tunnel

12. Main line for high-speed passenger trains
    Grande ligne - TGV de voyageurs
13. Main line for through freight trains
    Grande ligne - trains directs de marchandises
14. Shuttle arrival loop
    Boucle d'arrivée des navettes
15. Unloading platforms
    Quais de débarquement
16. Unloading overbridges
    Ponts d'accès à la route de sortie du terminal
17. Terminal exit road
    Route de sortie du terminal

18. Cars and coaches exit road
    Route de sortie des véhicules
    de tourisme et des autocars
19. HGV exit road
    Route de sortie des poids lourds
20. To A26 autoroute
    Vers l'autoroute A26
21. Development zone
    Zone de développement
22. Calais

Artist's impression of the Coquelles terminal
Dessin illustratif du terminal de Coquelles

1. Access road from A20 and M20
   Route d'accès au terminal à partir de la route
   nationale A20 et de l'autoroute M20
2. Toll booths
   Postes de péage
3. Terminal building
   Bâtiment du terminal
4. Control centre
   Centre de contrôle
5. Heavy goods vehicles (HGV) frontier controls
   Contrôles frontaliers pour les poids lourds
6. Frontier controls
   Contrôles frontaliers
7. Allocation area
   Zone d'affectation pour les véhicules destinés
   à l'embarquement
8. Loading overbridges
   Ponts d'accès pour les véhicules destinés à
   l'embarquement
9. Loading platforms
   Quais d'embarquement
10. Tunnel portal
    Portail du tunnel
11. Folkestone
12. M20 motorway
    Autoroute M20
13. Shuttle maintenance and stabling area
    Aire de garage et d'entretien des navettes

14. Shuttle arrival loop from France
    Boucle d'arrivée pour les navettes en provenance
    de France
15. Shuttle loop tunnel to platform area
    Tunnel en boucle pour les navettes en direction
    de la zone des quais
16. Unloading platforms
    Quais d'embarquement
17. Unloading overbridges
    Ponts d'accès à la route de sortie du terminal
18. Terminal exit road leading to A20 and M20
    Route de sortie du terminal en direction de la
    route nationale A20 et de l'autoroute M20
19. Main line for through trains operated by the
    national railways
    Grande ligne - trains directs exploités par les
    chemins de fer nationaux

Artist's impression of the Folkestone terminal
Dessin illustratif du terminal de Folkestone

The first breakthrough by a TBM occurred on 27 April 1989 when the French landward service tunnel TBM emerged at the portal at Beussingue

CHAPTER 8

BREAKTHROUGH

HOW *DID* THEY MEET IN THE MIDDLE?

To the often-asked question, 'Will the French and British tunnels meet in the middle?' the strict answer was, 'No, and something will have gone very wrong if they do'. If the two TBMs boring from each end of the service tunnel had proceeded inexorably along their laser-aligned paths until they came face to face, the result would have been a beautiful, 50-kilometre tunnel totally blocked by 500 metres of expensive machinery face to face, with tungsten-carbide cutting teeth literally gnashing at each other.

The engineers always knew that special arrangements would have to be made for the last few metres of boring and that those arrangements would probably entail at least one of the TBMs being left underground. The various options were kept open for as long as possible because there were a number of imponderables to be taken into account. What tunnelling speed would the French reach once they had cleared the fissured ground at the Calais end? Would the English crews be on schedule? Where, exactly, would the meeting-point be? What would be the nature of the ground at that point? The undersea 'sacrifice' of a TBM is not as wasteful as it might appear at first sight. These TBMs will have bored much further than many TBMs are designed to do in their entire working life. What is more, the cost of dismantling them underground and bringing them, piece by piece, along the tunnels to the surface is greater than the scrap value of the machines, since ground conditions and diameters vary so much. Over a year after completion of the land drive service tunnels in France and England, the redundant TBMs still have not been disposed of.

THE FATE OF THE TBMS

Three options were considered to deal with the TBMs at their meeting point. The first was to turn the heavy cutting heads of both machines off to left and right and concrete them in, thus entombing them for all time. The second alternative was to abandon only the British machine in this way. The French TBM, 'Brigitte', would then be dismantled and taken back to Sangatte. The third option also involved removing 'Brigitte' in the same way but the British machine, instead of boring its way in a curve towards its final resting place, would be rolled into a chamber excavated beside it and, again, concreted in.

In the end, ground conditions and speed of construction made the second option the favourite. It provided a means for tunnellers to gain access between the TBMs with the minimum of delay, an important safety consideration which also made it easier and safer to carry out non-TBM excavation.

THE LAST FEW METRES

As the moment of breakthrough drew nearer, tension throughout the project, from top to bottom, was very high. This was the focal point of the whole engineering enterprise, a few seconds of history in the making. This was the occasion at which all the service tunnel crews wanted to be present, although they knew that only the duty shift on the day would actually witness the first junction. It was breakthrough that fired the imagination of everyone connected with the fixed link and of large sections of the general public. Yet, as Colin Kirkland points out, imagination can leap ahead of reality:

The first breakthrough on the project was that of the service tunnel through Castle Hill, Folkestone, excavated by roadheader, on 17 April 1989

Breakthrough of the south running tunnel on 9 August 1989 through Castle Hill which, like all the Castle Hill tunnels, was excavated using the New Austrian Tunnelling Method (see page 83)

It's always exciting for tunnellers but it's frequently not as visually exciting as the average man in the street would think. He imagines, I suppose, hacking away at a huge wall of rock and seeing it collapse and there's the other chap standing the other side. That happens extremely rarely, and, of course, you couldn't allow it to happen under the sea. When you are 40 metres below the sea-bed and there are 60 metres of water above that, if the tunnel fell in the celebration would get rather dampened! So it's a great buzz but it's often visually pretty mundane.

What actually happened was this. The two TBMs came to a halt about 100 metres away from each other, with the head of the French machine precisely at the point of final breakthrough. Then, very cautiously, a hole, 4 centimetres in diameter, was bored through the final barrier of chalk marl to make what was technically the first undersea contact between Britain and France. To everyone's relief and satisfaction this final location check revealed that the margin of alignment error between the two tunnels was only a few centimetres. The British TBM was then driven on a curve to end up by the side of the French TBM but off the line of the service tunnel. The section of tunnel behind the cutting head was temporarily lined using steel ribs and timber logging rather than the pre-cast concrete normally used, since this section of lining would subsequently be replaced by cast-iron linings on the final line of the service tunnel. It was at this point that the hand-tunnellers came into their own. After all the advanced technology and massive machinery that had brought the two ends of the fixed link to this point, it was human muscles, hand-held and powered tools and shovels that finished the job. First of all, a hand-dug heading approximately 1 metre wide by 2 metres high was dug between the head of the French TBM and the rear of the British machine. While the French TBM was being dismantled within its outside shield and

Breakthrough of the French south landward running tunnel TBM on 18 December 1989. Owing to the relatively short length of the French landward tunnels (3.2 kilometres, compared with 8.1 kilometres on the UK side), a single TBM was used to bore both the north and south running tunnels

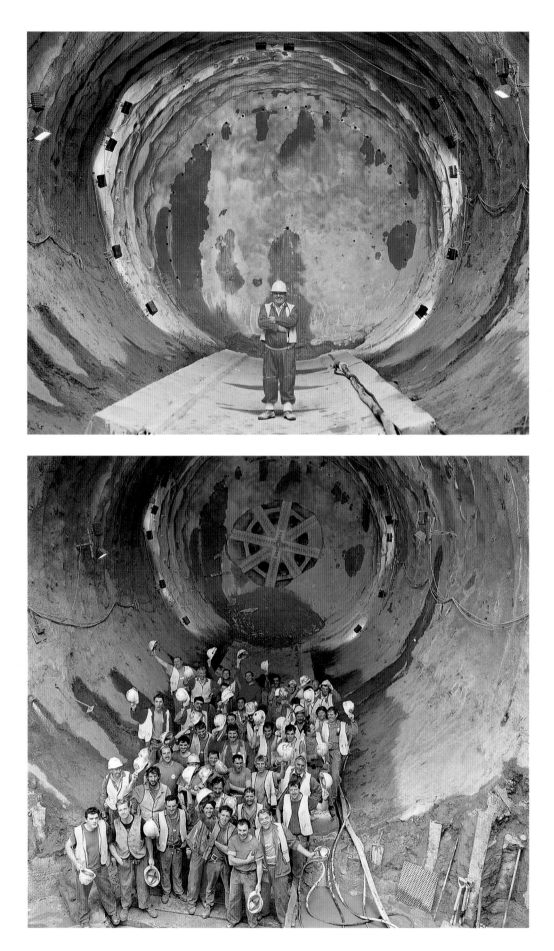

Moments before (top) and immediately after (left) the breakthrough of the UK north landward running tunnel on 11 September 1990. This was one of five TBM drives completed between September and December; the others were the UK south landward running tunnel, the French north landward running tunnel, and the French and UK seaward service tunnels.

The crossovers (four in total, of which two are undersea) will enable the shuttles and through trains to switch between the two running tunnels for maintenance and other operational reasons. The cavern excavated for the UK undersea crossover is 163 metres long, over 21 metres wide and 15 metres high, and is some 7.8 kilometres out from the Shakespeare Cliff Site.

The UK undersea crossover cavern (left) showing the `eyes' through which the two seaward running tunnel TBMs entered the cavern on their way towards France

The UK undersea crossover cavern was excavated using roadheaders and trimmed by hand (above left). Approximately 49,000 cubic metres of chalk were removed during the construction of this crossover.

Hand trimming in the crossover cavern (above right). Note the guidance system on the wall!

The north seaward running tunnel TBM broke into the UK undersea crossover cavern (right) on 27 August 1990. Less than a month later it was followed by its companion in the south seaward running tunnel.

Workers await the breakthrough of the south seaward running tunnel TBM into the UK undersea crossover cavern

taken back piece by piece to Sangatte, the final straight connection between the tunnels could be made from the British side. Since the TBM had already been parked to one side, this final connection had to be made by using a mechanical excavator called a road header and by erecting cast-iron linings, which are lighter to handle than concrete ones. At the same time, the sections of tunnel off line formed by the TBM were filled with concrete behind the cast-iron linings. Finally, the wall between the final portion of the British tunnel and the spot where the front of the French TBM had been was demolished.

IMPLICATIONS
Did breakthrough take place in France or in

Britain? It is not an idle question, nor as easy to answer as it might at first sight appear. The frontier between the two nations lies along the mid-point of the Channel, and that applies as much to the region beneath the sea-bed as to the waters above. Since the British service tunnel was longer than its French counterpart, there would come a moment when, strictly speaking, the crews from Shakespeare Cliff encroached upon French territory, with all kinds of far-reaching legal implications.

These potential problems were anticipated in the Treaty of Canterbury, which provides for the tunnel team to take their own law with them until an 'effective connection' is made. Much then depends on the meaning of 'effective

connection'. If it is defined as the moment of initial breakthrough, a number of practical problems would arise. For example, French Value Added Tax would be payable on British material and equipment installed beyond the frontier, and British workers would be subject to French health and safety legislation. The resulting uncertainty and unfamiliarity could have caused real safety risks. If, as expected, the next machine (in running tunnel north) to cross the frontier were French, things would have been even more complex, with teams in adjacent tunnels subject to different laws! Fortunately, the two governments reached a common-sense solution: 'effective connection' would not be deemed to have happened until the start of the

The UK south seaward running tunnel TBM passing through the UK undersea crossover cavern

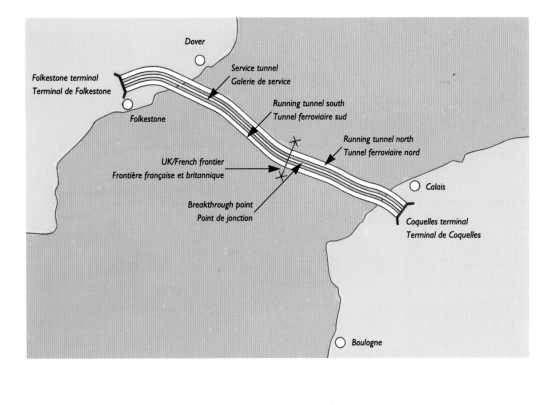

Dover

Folkestone terminal
Terminal de Folkestone

Folkestone

Service tunnel
Galerie de service

Running tunnel south
Tunnel ferroviaire sud

Running tunnel north
Tunnel ferroviaire nord

UK/French frontier
Frontière française et britannique

Breakthrough point
Point de jonction

Calais

Coquelles terminal
Terminal de Coquelles

Boulogne

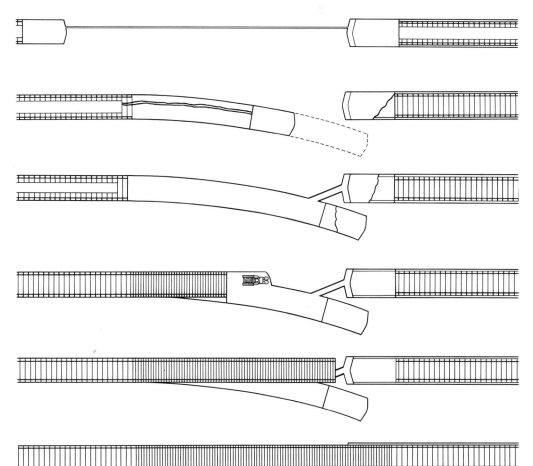

How the Service Tunnel Breakthrough Happened

The French TBM stopped at a predetermined location and the UK TBM turned off to one side to end up beside the French one. A small heading or hand dug access about 2 metres high and 1 metre wide was constructed between the two tunnels. The UK TBM cutting head was left in place and concreted up, while the French TBM was dismantled and removed to Sangatte. The final connection tunnel was then completed working from the UK side, using the New Austrian Tunnelling Method with excavation carried out by a roadheader machine before cast iron linings were installed. The last 'wall' of the tunnel was excavated near where the French TBM stopped.

An engineer on the UK seaward service tunnel TBM checks the figures relating to its alignment (above)

French tunnel workers sign the final tunnel lining ring of their seaward service tunnel (right)

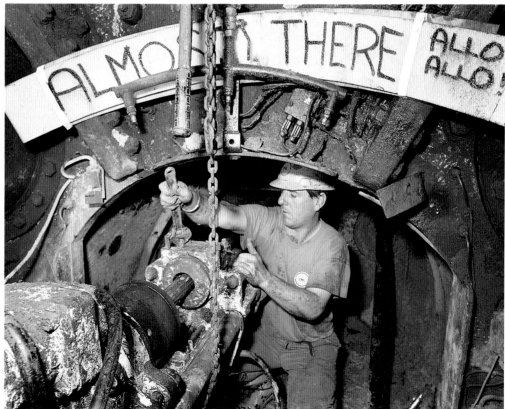

The probe drill on the UK seaward service tunnel TBM (above). Normally, the probe drill is used to monitor the condition of the ground in advance of the TBM's drive but here it is boring into the face towards the French service tunnel in order to establish the alignment of the two TBMs.

`Almost there' (right). When the probing started the two TBMs were approximately 110 metres apart.

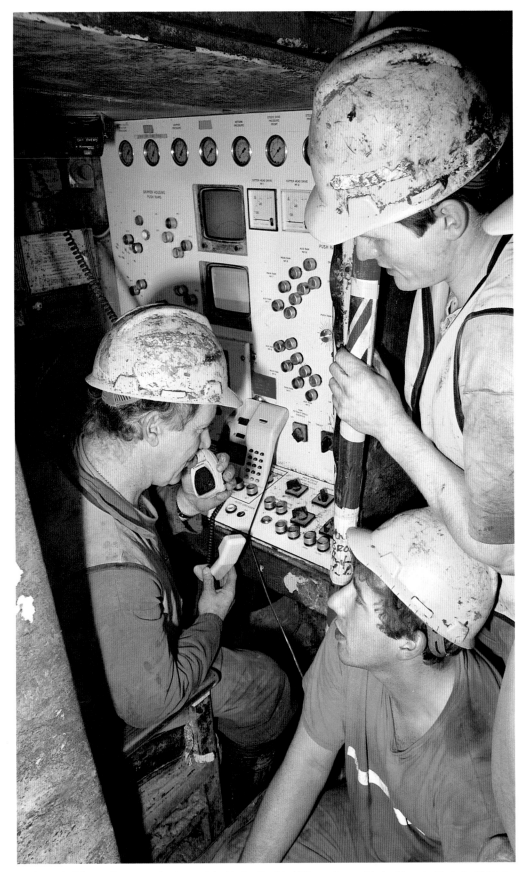

Workers in the UK seaward service tunnel gather round the driver's cabin of the TBM to await news of the breakthrough of the probe drill into the French seaward service tunnel

commission phase of the project, planned for the end of 1991.

Despite this interpretation, it was still necessary to introduce frontier controls from the moment of initial breakthrough. For practical purposes, the frontier control authorities consider the construction site and tunnels as one 'foreign territory', so that, effectively, all tunnel workers go 'abroad' every time they go to work.

LIVING WITH THE CHANNEL TUNNEL

So, the United Kingdom and the Continent are linked. By a kind of reverse-birth process an umbilical cord has been created and brought a new entity into being. Britain has become what one French newspaper called *une presqu'isle*, an 'almost island'. There are still the two running tunnels to be completed, as well as the terminals and road and rail links. Track has to be laid, overhead catenary, power, lighting, signalling, cables, ventilation equipment, fire and drainage systems all have to be installed and tested, and the shuttles have to be built and commissioned. But the turning point has been reached. The project, begun less than four years previously, has passed its critical point. The voices of scepticism and doom are falling silent.

The major obstacle to the establishment of a workable fixed link, however, was never those few kilometres of chalk and chalk marl. The real hurdle which Eurotunnel must yet surmount is lodged in the hearts and minds of the general public across Europe, but especially in Britain, where many people remain sceptical. To be viable the Channel Tunnel needs paying customers - lots of them. Breakthrough was an engineering triumph. It pointed the way to a new kind of Europe. Is it the kind of Europe that most citizens want? If so, Eurotunnel must convince them that the Tunnel is the vehicle for a fast, safe and reliable passenger and freight service. Public attention now shifts from the Tunnel to what will pass through it.

First of all there is concern about what should not pass through it. Eurotunnel is required to ensure that the creation and operation of the Tunnel will not increase the risk of rabies entering Britain. Obviously, Eurotunnel cannot

Expectant faces as workers in the UK seaward service tunnel wait for news of the breakthrough of the probe drill into the French seaward service tunnel

be held responsible for animals being smuggled across the Channel in small boats or in cars on the ferries. To minimize the risk of animals being smuggled in cars through the Tunnel, the frontier control services will maintain similar checks to those used on ferries at the moment. An additional benefit of people staying with their cars on the shuttles is that passengers in adjacent cars may well detect animals in cars. Customs people say that seeing an animal being smuggled is one of the few occasions when the British public is prepared to tell tales on their neighbours.

In addition, the tunnels will be kept clean and rubbish-free, and it will not be possible to throw food out of the train windows or discharge lavatories on to the tracks. Clean tunnels reduce the incentive for animals to enter in search of food. Lastly, electric grids within the tunnels and animal-proof fencing will further deter animals from entering. All this work has been prepared as a result of tests and studies undertaken by many organizations in the UK and France, including the Rabies Study Centre at Nancy.

These precautions relate to only one of the multitudinous aspects of public safety with which Eurotunnel is concerned and on which it has to satisfy British and French regulations. The Franco-British Intergovernmental Commission (IGC) was set up under the Anglo-French treaty to approve the design and construction of the system, particularly those elements relating to safety. IGC approvals have to be received before Eurotunnel can start operations. The ICG is advised where appropriate by the Safety Authority, which comprises senior UK and French officials highly qualified in the fields of rail safety, civil engineering, health and safety at work and fire detection and prevention. Confidence and safety measures are considered very important for any transportation system to be a commercial success. Eurotunnel aims for a comprehensive and integrated approach to safety both in the design of the system and its operation.

For example, two separate running tunnels are safer than one tunnel with two tracks, and the service tunnel provides emergency access and

The probe drill emerged in the French seaward service tunnel at 1930 hours GMT on 30 October 1990 and established that, after a combined tunnelling distance of approximately 38 kilometres, the two TBMs were only a few centimetres off line

Workers in the UK seaward service tunnel celebrate the breakthrough of the probe drill. This historic connection below the seabed linked Britain and continental Europe for the first time since the last Ice Age.

ventilation. The shuttles are designed to incorporate non-combustible material wherever possible. The shuttle structure has a 35-minute fire existence, and incorporates a wide range of fire detectors and suppression systems as well as a comprehensive control signalling and communication system. The intention is, by demonstrating the highest standards and a professional approach, to give future customers every confidence in using the system.

When completed, the Channel Tunnel will be a very sophisticated hole in the ground connecting the road and rail infrastructures of Britain and France. The Tunnel is just one of the many developments (and not necessarily the most significant) that are currently creating a new political, social and economic map of Europe. Only time will tell its impact, both on its

immediate hinterland in Nord-Pas-de-Calais and Kent and on the wider pattern of trade and travel in Western Europe. But this first physical land link between Britain and the Continent for some 12,000 years has the potential to provide the practical as well as the psychological connection that will encourage Britain, economically and politically, to play its full part in the Europe of the 1990s and beyond into the twenty-first century.

ANDRÉ BÉNARD

La Défense, sometimes referred to as 'Manhattan-sur-Seine', is a spectacular development of skyscrapers and audaciously modern commercial buildings on the edge of Paris. It is in one of these high-rise temples to human enterprise that you will – if you are lucky

– track down André Bénard, Chairman of Eurotunnel. It is a fitting habitat, for Monsieur Bénard shares with many of his countrymen a commitment to technical progress. For him the Tunnel is an inevitable development.
I didn't have to be convinced that there needed to be a fixed link between the UK and France. I've lived in the UK for many years. I've been through the war and fought with the British. I had 44 years with Shell. So my relationship with the UK is a very longstanding one. I've never had any doubt that you needed to have a fixed link.

André Bénard joined Shell in France in 1946. He served as Chairman and Chief Executive of the French affiliate of Shell from 1966 to 1970 before moving to become Managing Director of Royal Dutch Shell. So he brought years of experience and considerable managerial skill

Dismantling the French seaward service tunnel TBM, its work completed. The dismembered TBM was taken back along the tunnel in the wagons previously used to carry the spoil created by the machine.

Near the point of breakthrough the French TBM stopped while the UK TBM veered to the right and finished beside the French TBM. The relatively tight curve of this manoeuvre (a radius of approximately 200 metres) required the removal of various pieces of the UK TBM, as shown here.

To achieve stage two of breakthrough a heading, or hand-dug adit, was created between the two tunnels from both the UK side and the French side, shown here

when he joined Eurotunnel in 1986. He found it a very different animal.

I had worked for a company which has probably the best teams available anywhere in the world for operating a large project. We had the best legal advice available, the best technical advice available, the best financial advice available, and a company culture. Eurotunnel is not at all like that. When I joined Eurotunnel, it was just a name which had been filed by contractors and bankers. Eurotunnel didn't exist as an entity. So we were faced with quite a challenge.

Few leading businessmen have Monsieur Bénard's facility in spanning two cultures. His ability to understand the different attitudes of French and British participants in the project has been vital in creating an effective team and in obtaining the backing of the financial institutions in Paris. The task has never been an easy one. Why did he persevere with it? It was partly the challenge; partly the conviction that the fixed link makes sound business sense; but, above all, it was his philosophical dedication to the European idea.

What ultimately matters, he suggests, is not commerce but quality of life. Europeans cannot go on living in closed national boxes and the Tunnel is one tool for opening the lids. 'The Tunnel led to the decision to develop the TGV nord. The TGV nord led to the decision to build a European fast-train network. The fast-train will undoubtedly lead to greater exchanges between nations than anything else you can think of.' This French Anglophile believes that Britain has much to give continental Europe – 'wisdom, reason, a sense of balance, humanistic philosophy' - and that the Tunnel can be a conduit, not just for vehicles and people, but for ideas.

ALASTAIR MORTON

It's taking Britain a century to re-orientate from being the heart of a maritime overseas empire with its back to Europe to being part of an integrated Europe. The process had better speed up a bit, or by the time it's finished we will be too late for the party.

Alastair Morton, Deputy Chairman and Chief Executive of Eurotunnel, is tough, outspoken, impatient with bureaucracy and blinkered thinking. His commercial career is impressive.

At this point the French and UK service tunnels were only 3 metres apart, enabling the workers in the two tunnels to talk to each other (above) through the hole bored to ensure the correct alignment of the heading, or to pass objects such as this one franc coin to each other (below)

After a succession of high-powered jobs in South Africa, the USA and Britain, he became the first Managing Director of the British National Oil Corporation in 1976 and the Chief Executive of Guinness Peat financial services group in 1982. He was, thus, no stranger to firm decision-making and boardroom battles. What was new to him was the high profile of his new position. Because of public interest in the Channel Tunnel, difficulties which would normally have remained internal matters have attracted press comment.

Alastair Morton has, to some extent, been able to turn all this publicity to advantage. He has spent much of his time campaigning for a vastly improved British transport system which, together with the Tunnel, will integrate Britain effectively with the commercial life of Europe. To him it is crystal clear that history underlines this lesson.

You had the canals, you had the railways, you had the motor highways at the beginning of this century. Where they ran was where the towns and industries grew up. Much urban development followed transport facilities rather than pre-dating them. Now, Europe is becoming a region of the world to match the Asia-Pacific Basin and to match North America; a very large and relatively integrated economic region that will run from somewhere around Warsaw and Budapest to Swansea, Liverpool and Glasgow.

He regards preaching the gospel of 'Britain in Europe' not as a hobby but as 'a significant part of my job'. He believes that people, especially in Kent, are coming round to realizing the importance of efficient transport links between Britain and her cross-Channel trading partners but he is under no illusion that the mere existence of the Tunnel will complete the re-evaluation process. 'It's an enormous change and you can't expect it to happen in twenty years from 1973.' But time, he says, is running out fast.

Alastair Morton has also been particularly associated with the campaign to raise the funds for the project.

It was no coincidence that the underwriting of the additional equity of £566 million raised in 1990 was completed within 3 days of the first undersea contract on October 30, 1990. *André Bénard and I had thought since the autumn of 1989 that the best moment to launch the rights issue would be at the time of breakthrough - more than a year ahead. The launch also followed the signature of the revised credit agreements on 25 October 1990 that increased the commitment of the bank syndicate by £1.8 billion and added £300 million from the European Investment Bank. In total we increased the available funding by £2.666 billion to £8.767 billion in aggregate. Even American audiences find the idea of $17 billion private sector project financing hard to digest. Both the tunnelling and financial teams did a magnificent job during 1990.*

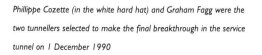

Phillippe Cozette (in the white hard hat) and Graham Fagg were the two tunnellers selected to make the final breakthrough in the service tunnel on 1 December 1990

The breakthrough achieved, Phillippe Cozette and Graham Fagg exchange flags and congratulations

Above ground, French tunnellers celebrate their colleagues' success under the seabed

Toasting success, moments after the service tunnel breakthrough

ACKNOWLEDGEMENTS

The author and publishers would like to thank all those connected with the Channel Tunnel project

who have assisted in the preparation of this book.

The description of the system contained in this book is that envisaged at the time of printing. The design,

development and construction of the system and the planning for its operation are combining, and are subject to

modification before the planned commencement of operations in 1993.

All drawings and diagrams included in the book are illustrative only.

ILLUSTRATION CREDITS AND SOURCES

The illustrations in the book have been provided by the following sources and copyright holders. No photographs or

artwork may be reproduced without the express permission of the copyright holders,

The Channel Tunnel Group Ltd, and Random Century.

Page 1 P. Perrin/Sygma; 8 The Telegraph Colour Library; 12 Island Artists; 16/17 (top left) Robert Kirkman; (top right)

Sealink British Ferries; (bottom) John Laing Construction Ltd; 20/21 The Channel Tunnel Group Ltd; 22/23 The

Channel Tunnel Group Ltd; 24 The Channel Tunnel Group Ltd; 25 The Channel Tunnel Group Ltd; 28/29 The

Channel Tunnel Group Ltd; 29 The Channel Tunnel Group Ltd; 32/33 The Channel Tunnel Group Ltd; 35 P.

Perrin/Sygma; 36 The International Stock Exchange Photo Library; 37 The Channel Tunnel Group Ltd; P.

Perrin/Sygma; 71 (right) P. Perrin/Sygma; 80\81 The Channel Tunnel Group Ltd; 91 White Cliffs Countryside Project;

94 (top) The Channel Tunnel Group Ltd; (centre) Canterbury Archaeological Trust; 104\105 Newington Frith in

association with The Society of Antiquaries of London; 108\109 SNCF; 110 (top) Ville de Lille\Daniel Rapaich;

(bottom) Association TGV Amiens-Picardie-Normandie; 111 Renault France; 112\113 The Telegraph Colour Library;

114 (top) Syndication International; 115 (top right) P&O; (bottom) Hoverspeed; 116 EEC Library; (bottom) The

Channel Tunnel Group Ltd; 118\119 British Railways Board.

All other photographs taken by QA Photos Ltd, copyright The Channel Tunnel Group Ltd

ARTWORK CREDITS

Page 18/19 Arthur Phillips; 38/39 Kevin Jones Associates; 42/43 David Eaton/The Organisation; 44/45 Arthur Phillips;

66/67 Kevin Jones Associates; 118/119 Neil Gower;

130 (colour artwork) David Eaton/The Organisation; 130 (black and white artwork) Arthur Phillips

Book design by the Senate

INDEX

access shafts 9, *32, 47-9, 48-50, 71*

air routes 12

Albert, Prince 24

Amiens *108,* 109

Ancient Monuments and
  Archaeological Areas Act (1979) 91

archaeology 91, 96, 102

Ashford *83,* 114, 115

Association du tunnel sous-
  marin (1870s) 41

Baeckeroot, Christian 108-9

Balfour Beatty Construction Ltd 32

banks 12, 15, 32, 36-7

Banque Indosuez 32

Banque Nationale de Paris 32

Bateman, John 29

Beaumont, Colonel, and B.
  tunnel 39, 41

Beaumont and English boring
  machine *22-3, 26-7*

Belgium 114, 115, *116*

Bénard, André 107, 135-7, 139

Bermingham, Peter 45, 60-3

Biggins Wood 97, 102

Bonaparte, Napoleon 21, 23

Boulogne 78, 109, 114

Bourse, La *36*

Bouygues SA 32

'Brigitte' (TBM) 50-3, *121*

British Rail (BR) 12, 17, 37, *84,* 115,
  see also rail links; trains

Brunton drilling machine *25*

Calais 78, 109, 114

Camp du Drap d'Or 79, 102, *105*

Canterbury Archaeological Trust 102

Canterbury, Treaty of (1986) *13,* 31

Castle, Barbara 11

Castle Hill 83, *86*

'Cathérine' (TBM) 53

Channel Expressway project 16, *16-17*

Channel Tunnel Act (1987) 31,
  93, 94, 96, 99, 115

Channel Tunnel Company (1872) 24, 41

Channel Tunnel Group Ltd -
  France-Manche SA (CTG-FM)
  16, 31-2, 34, 36

Churchill, Winston 26

Complaints Commissioner 100

concrete and linings 41, 49,
  58-60, 60, 83, *86, 88-9*

conservation projects 79, 91,
  94-6, *95, 97-8,* 99, 102

Coquelles terminal 41, 77-9,
  77, 78, 80, 84-5, 118

Costain Civil Engineering Ltd 32

costs 36, 37, 50

Countryside Commission 94

Cozette, Phillippe *139, 140*

Crédit Lyonnais 32

cross passages 32, 49, *66-7*

crossovers 49, *126-9*

Cuffley, Linda 114

Culbard, Elisabeth 101

cut-and-cover technique 83, 86

Delebarre, Michel 109

Denman, Davie 63

Desmarest, Nicolas 21

Directive on Environmental
  Impact Assessment 96

Dollands Moor *84*

Dover *51, 113,* 114

drainage system 67, 69

Dumez SA 32

Dunkirk *50,* 78, 109, 114

EC see European Communities

engineering see technology

Etchinghill Escarpment 99

Eurobridge project 16, *16*

'Europa' (TBM) 53

European Commission see European
  Communities

European Communities (EC) 10, 12,
  17, 91, 96, 114, *114*

Euroroute project 15-16, *17*

exhibitions and information
  centres 37, *100,* 101-2, *101, 103*

Fagg, Graham *139, 140*

Farthingloe village 63, *70-1*

ferries 12, 21, 109, 112, *113*

Field of Cloth of Gold (1520) 102,
  104-5

Figuier, Louis 21

finance 12, 15, 32, 36-7,
  see also costs

fire precautions 67, 72

First World War project 26

Folkestone: information
  centre and exhibition 100,
  100, 101-2, *103;* terminal
  41, 76-7, *77-84, 79, 81, 82,*
  86, 93, 119; tourism 114

Fond Pignon 65, *74*

France-Manche SA see CTG-FM

Francis I, King 102, *105*

Frogholt 41

Gaulle, President Charles de 10

geology 39-41, *42-3,* 60, 78

Gleeson, John 45, 55

Global Positioning System 45

Goodwin Sands 83-101

government involvement 10-15,
  31, 36, 99, 115

Gueterbock, Tony 101

hand tools see roadheaders
  and hand tools

Hawkshaw, Sir John 24

health and safety 71-2

Henry VIII, King 102, *105*

Historic Buildings and
  Monuments Commission 94

Holywell Coombe 6, *86,* 96

house purchase 101

Intergovernmental Commission (IGC)
  133

Isle of Grain plant *58,* 60

Jenkins, Windsor 55

Kent County Council 115

Kent Impact Study 112-14

Kent Trust for Nature
  Conservation 99

Kershaw, Katharine 101

Kirkland, Colin 41, 94-6, *121*

landward tunnels 40, 49, *56-7,*
  61, 68, 83

legislation 31, 36, 91 see also Channel
  Tunnel Act